Illustrated Sketches of Death Valley

Also from Westphalia Press

westphaliapress.org

Illustrated Sketches of Death Valley

and Other Borax Deserts of the Pacific Coast

by John R. Spears

WESTPHALIA PRESS
An imprint of Policy Studies Organization

Illustrated Sketches of Death Valley and Other Borax Deserts of the Pacific Coast

Westphalia Press
An imprint of Policy Studies Organization
1527 New Hampshire Ave., NW
Washington, D.C. 20036
info@ipsonet.org

ISBN-13: 978-1-63391-541-1
ISBN-10: 1-63391-541-7

Cover design by Jeffrey Barnes:
jbarnesbook.design

Daniel Gutierrez-Sandoval, Executive Director
PSO and Westphalia Press

Updated material and comments on this edition
can be found at the Westphalia Press website:
www.westphaliapress.org

VIEW ON THE MOJAVE DESERT.

ILLUSTRATED SKETCHES

OF

DEATH VALLEY

AND OTHER

BORAX DESERTS OF THE PACIFIC COAST.

BY

JOHN R. SPEARS.

CHICAGO AND NEW YORK:

RAND, McNALLY & COMPANY, PUBLISHERS.

1892.

The author begs leave to dedicate these sketches to the Copy Editors of the *Sun*. Having been accustomed to receive and accept praise for articles called creditable, because crudities had been eliminated and bright hits added by the gentlemen of the blue pencil, he desires now to share the regret the reader must feel because this work did not, as the rest has done, pass under their discriminating eyes.

A FAIR WARNING.

Early in the year 1891, while a series of sketches made during a trip to the Cryolite mine in Greenland was running through the *Sun*, an associate on the staff of that paper, Mr. S. T. Mather, came to me and said:

"Let me tell you where to go to get another lot of sketches; go out among the borax deserts of the Pacific Coast."

Before the matter could be properly considered I was sent by the chief to Central America, and by the time I had returned home the deserts had been entirely forgotten. Afterward, while making some purchases in a grocery store, I noticed that the label on a package of borax was more or less (chiefly less) beautified by a picture of two wagons drawn by a string of mules that stretched out apparently for half a mile over a boundless plain.

It is a poor soul that does not feel interested in even one mule with its driver, but to see a half-mile string of desert mules (possibly a new variety) with their driver, that would be a spectacle full of such stirring and hilarious possibilities. But to continue: Above and below the picture appeared the legend "Borax from the Deserts." That reminded me of what my friend had said, and I began to investigate the matter of a journey to the Pacific Coast deserts. The men of whom I made inquiry told me, among many other things:

That it was a land of myths and mirages;

That it was the abiding place of a novel race of Arabs;

That it had originated an entirely unique species of tramp;

That it was "the home of the chahwalla;"

That the mules were a reality.

The discoveries of the gold and silver deposits and lodes of the Pacific Coast have received so much attention at the hands of writers, that pictures of life in the mining camps are about as familiar to the reader of the East as to the resident of the West. We citizens of the metropolis know all about Pickhandle Gulch and its horde of men with pans and cradles; we have seen, so to speak, sluices and spurting streams that tear down solid rocks; we are at home among the shafts, the tunnels, the stopes of the San Francisco Belle or the Consolidated Virginia. These pictures of life, aside from the glamour which a precious metal throws over everything it is associated with, have always been found fascinating because of the stirring of rude characters depicted.

Since this was true, might not the undescribed life on the Pacific Coast deserts be of human interest? Would anybody want to read about the American Arab, and the desert tramp, and the chahwalla, and the mule? If not, why not?

So I made the sketches, and here they are. *Sauve qui peut*

J. R. S.

INSCRIPTION ROCK.

SPANISH BAYONET.

LIST OF ILLUSTRATIONS.

DEATH VALLEY.

CHAPTER I.

THE STORY OF DEATH VALLEY.

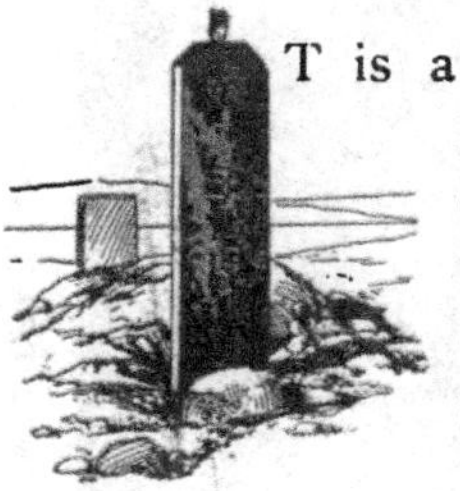

IT is a gruesome story of a rugged country. More than a score of lives were lost in a day when the valley was christened, and its history from that day to this has been one of hardship, peril, and death, with rarely aught to relieve its harshness.

It is a story, too, of apparent paradoxes and of wonders. Nature, if unkind in a way, has been lavish in her gifts to this desert pit. Well has the valley been named, and yet for more than half of the year it is one of the healthiest spots on the Pacific Coast. It is a place where rain-storms are well nigh unknown, and yet one where the effects of cloud-bursts are almost unparalleled. It is the hottest spot on earth, and yet ice often forms there. It is a place where the air becomes so arid that men have died through lack of moisture when abundant water was at hand, and yet the stopping place of hundreds of ducks, geese, and other migrating water-fowl. It is a region where the beds of lakes are found on the pointed peaks of mountains. It is a region where a mountain system of the most gorgeous-colored rocks is known as the Funeral Range. It is a rent in the earth, the bottom of which, in spite of the washings of

LOOKING INTO DEATH VALLEY.

centuries, is probably deeper below the level of the sea than that of any other valley in the world.

Surely the story of Death Valley should have been preserved, but, unfortunately, although scores of articles have appeared in print on the subject, they have usually been imaginative, and even so thorough and accurate a work as "Bancroft's Pacific Coast History" contains only a brief and unsatisfactory reference to the matter. The history of Death Valley is found only in tradition. As I gathered it, here it is:

In the year 1850 the number of parties of emigrants bound to California from the Eastern States was so great that their trains of wagons formed what may be called almost a continuous procession from the Missouri River to Salt Lake City. It is said that many a lone traveler, bound overland on horseback, with perhaps a single pack animal, found a hospitable welcome every night of his trip at the camp of some party of emigrants, and yet never stopped twice with the same party until he reached the Mormon settlement. The similarity to a procession would, indeed, have been found west of Salt Lake, but for the fact that the parties commonly divided at that point, some going on by the route which was afterward followed by the Central Pacific Railroad, while the rest struck down through Utah, Nevada, and Southern California, bound through the Cajon Pass for the regions of which Los Angeles was then, and is now, the metropolis.

There was a great variety among these emigrant outfits, but the majority of the parties had huge prairie schooners drawn by oxen. The outfit was one that had been found best for such journeys by explorers like Fremont and the early Santa Fé traders. Oxen were not only able to live on the grass that grew along the trail; they got fat on it. And in an emergency, such as an attack from the Indians in which animals were killed, the dead animals were by no means wholly wasted. It was easy and comforting to cure and eat dead oxen, but mules were different.

Among the parties in that long procession was one that was destined to give Death Valley its name. The facts we don't know about that party would make an interesting book, no doubt; but this much tradition tells: On reaching Salt Lake they struck off to the south, because the northern, or Truckee River, route had been traveled so much that feed and fuel (the land being a desert) were scarcer than to the south. There was nothing unusual about that move, however, for a good many parties did the same thing, traveling along the trails leading near the west bank of the Colorado River for a few hundred miles, and then striking across the desert, by the way of several well-known springs, to the Mojave River, that sinks in the sands of the Mojave Desert. But this party was not content with what seemed to them a roundabout route. They were bound to get to the land of gold a little ahead of the others, so somewhere in Nevada, perhaps near Duck Lake, but maybe farther south, near Clover Valley Cañon, they left the trail to follow the compass. The western part of Lincoln County, Nevada, is well cut up with trails in these days, and there are many springs to be found there around the Pahranagat Range. The emigrants had no trouble there. The lava beds south of the Kawich Valley naturally turned the party to the south still farther, and then, seeing the bare mountains before them, they got down into the Amargosa River valley and the vicinity of Ash Meadows.

They were still safe. They had found springs of water at such intervals that, with the aid of kegs and barrels, they had been able to keep their animals in condition, fit to work in spite of the heat, but their situation was rapidly becoming desperate. The country was becoming more rugged, the valleys were narrower, the mountains were precipitous, the cañons more obstructed with the debris of tornadoes and cloud-bursts. They now began to write in the sand, with abandoned equipages, the story of their accumulating misfortunes. Article after article of household furniture—everything not necessary for the immediate use of the party—was cast aside to lighten the loads, and the women as well as the men walked beside the wagons rather than burden the worn-out cattle.

Finally, the Funeral Mountains rose across their pathway, and with weary toil they followed up a torrent bed between two peaks, to find themselves, when on the summit, overlooking a deep and narrow valley, whose walls were more precipitous and rugged than any they had yet seen. In the glare of the sun the long, narrow salt marsh that winds down the center of the valley looked like a cooling river of water, but the emigrants, after their experience in crossing Nevada, were in no way deceived by the mirage. It but added to their anxiety and apprehensions, as they unyoked the cattle and prepared with ropes and chains to lower the wagons by hand down to the mesa at the foot of the ridge on which they were standing. It was a fearful task in that atmosphere. The men were now not only worn out by the long journey over the deserts they had crossed; they found the lassitude, due to the lack of moisture in the air, almost too much for even the will of a pioneer to overcome. But they succeeded at last in reaching the mesa, and there night overtook them. Then while a few searched without success for a spring of water, the rest, with prudent forethought, made an orderly camp, stretching out the chains before the wagon-tongues and putting the yokes in place across them, while the cattle were turned loose to graze.

Fires were made from the scant fuel of the desert—the grease-brush—supper was cooked and eaten with little or nothing to drink, and then all prepared for the most pitiful experience that comes to the traveler, the passing of a night in a dry camp—a camp without water—a camp in which the cattle bawl, the men toss about, and mothers with

DEATH VALLEY, NEAR EMIGRANTS' LAST CAMP.

breaking hearts vainly strive to soothe the little ones wailing for want of drink.

Wretched as was that last camp, its sufferings were but the prelude to the terrors of the coming day. With the first streak of light, the search for water and a pass into the snow-capped range in the west were sought for. It was a hurried search from the first, a search that under the smiting rays of the sun quickly became feverish and at last delirious. Abandoning camp and wagons in their frenzy, the party separated, and in groups spread out to the north and the south along the face of the Panamints, walking over sand so hot that even the desert Arab, inured to its terrors, wraps his boots in sacks when obliged to cross it at mid-day; over a marsh, covered with a crust through which the foot breaks to sink in corroding brine; climbing up gulches where the black rocks seared their hands and the stirring of the air was like a blast of flame.

There were thirty souls in that party, of whom perhaps a dozen got beyond the Panamints. Of this number, a man named Towne, with his wife and one or two others, reached the Argus Range, and camped there while they killed a couple of oxen and dried the meat, saving one animal that Mrs. Towne might ride it. Bones of men who had tried to follow them across the Slate Range, but were unable to do so, were afterward found by prospectors. Sidney P. Waite, another one of the party, eventually escaped, and in 1890 was living at San Bernardino, Cal. Last of all was a man named Bennett, and he it was who really gave the valley its notoriety.

Of the rest, several skeletons of men were found by Dr. S. G. George, while prospecting in 1860. The men had fallen down and died within 300 yards of a spring of good water. Their bones were buried where they fell. Where the rest lie, nobody knows.

But pitiful as was their fate, it would long since have been forgotten, and the valley to which their fate gave a name would have been still a land unknown, had not Bennett, some time after reaching a civilized community, asserted that he had found a ledge in which pure silver cropped out. It came about in this way: As he stumbled along in a cañon in the mountains west of the valley, he found a spring of water, and stopped to drink and rest. His life was saved, for the time at least, and here he remained until he had regained somewhat of his old-time strength. While sitting idly by the water, he broke off a bit of an exposed rock, and was surprised to find it of metallic weight. So, thinking it might be something of value, he put a small piece in his pocket, and eventually carried it to the settlements. There he got a gun that needed a front sight, and for the sake of making a memento of the metal-like stone he had carried, he asked a gunsmith to make the sight of it.

To the astonishment of everybody who knew the circumstance, the bit of rock was found to be silver, and thereat the story of the Gunsight lead was created, and floated up and down the coast aimlessly until the famous strikes made at Virginia City turned the mining world upside down and shook it. There never was such a time in the history of the nation as the Bonanza era. Men with costly outfits, and men with scant rations as well, wandered off wherever there was a mountain or hill to be found, their eyes forever on the rocks about them and their minds on a mirage of delights which a bonanza strike, ever to be made in the next cañon, would bring them. They were drunk with the thought, insane with their greed. What were the terrors of the Mojave Desert, or even of Death Valley itself, to men like that?

In May, 1860, ten years after the emigrant party had perished in Death Valley, Dr. Darwin French made up a party in Butte County, California, to go in search of the Gunsight lead. They traveled by the way of Visalia, the south fork of the Kern River, and Walker's Pass to Little Owen's Lake, and thence easterly across the head of the valley next west of Death Valley, and through a rocky pass to Death Valley itself. They had planned the route so well that they arrived in the valley at the very camp where the emigrants had passed their last night together. They found wagons, yokes, chains, guns, revolvers, cooking utensils, even the toys of little children, lying about as they had been left ten years before. No rain had fallen to wet them, no sand had covered them, and the passing Piute Arab, knowing the fate that had overtaken the party, had hurried away in superstitious terror.

THE OLD EMIGRANT'S OX-YOKE.

Some few things were gathered for relics, and tnese eventually found place in the State Museum in San Francisco. A creek running into Death Valley from the Funeral Mountains was found and named, as told elsewhere, Furnace Creek, because they found old lead-smelting furnaces there which the Mormons had used during the troubles of 1857. But the season was not propitious for Dr. French's party. The hot weather was at hand, and in spite of abundant water, they found it expedient to get out quickly to avoid the fate of the emigrants.

In October of the same year (1860) came Dr. S. G. George, with others, in quest of the Gunsight lead. They followed the same route as Dr. Darwin French's party had, and reaching the old emigrant camp, found it still undisturbed. They had come well prepared, and prospected the mountains very thoroughly in all directions. They found excellent water in several places, by digging wells. The emigrants might have got it had they known. Following an Indian trail through the Panamints they reached what is known as Wild Rose Spring, and on Christmas day discovered a mine of antimony, which they named the Christmas Gift. But they did not find the Gunsight, nor did their Christmas gift make them rich.

This party found quite a different climate from that depicted by the emigrants and the previous visitors, and they were rather inclined to scout the idea of Death Valley being very deadly. Ducks and other birds abounded, while jack-rabbits and cotton-tails bobbed about. There was nothing particularly bad about Death Valley, so far as they could see, and, therefore, widely-varying stories about the region went up and down the coast.

Meantime, and as long ago as 1856, a gold vein of low-grade ore had been found near the Amargosa River bed, east of Death Valley, and in 1861 a party of eight Mexicans went to it and started in to work it. They got on fairly well for a time, but when they had built a small mill the Piutes came along in force, killed all the men, and burned the mill. This is referred to only to illustrate another danger that hung over Death Valley. The hardy prospectors had to pack rifles, and go in parties and keep watch for the signs of the treacherous Piutes.

In March, 1861, when the Mexicans were starting in at Amargosa, Mr. Hugh McCormack went to Death Valley. He discovered a spring, known as McCormack's Wells on some of the maps of California, and near where Mesquite Well now is. At the lower end of the valley he found the skeleton of a woman, with part of an old calico dress wrapped about it. It had been buried in a shallow grave, but the wind had uncovered it.

A month later (April, 1861), when the nation was aflame with the first powder burned in the great war, Mr. W. T. Henderson entered Death Valley, looking, as the rest had done, for the Gunsight lead. He did not find it, but he did what no man had done before—he climbed to the top of the highest of the Panamints, and standing there looked off over such a landscape as can be seen nowhere else on earth. To the west lay the Slate, the Argus, and, blue with the distance, the Sierra mountains. To the south rose Pilot Butte, the Calicos, and far away the San Bernardino Range. To the north were the snowy White Mountains, while to the east, beyond the Funerals, were the Ivanwatch, the Granite, and range after range that had never been named. Between them all lay the valleys, yellow with sand and grease-bush, spotted with black lava buttes and brightened with the beds of soda, salt, and borax, that gleamed snow-white to the eye, or turned to mirage lakes, with dancing waters and leafy borders, according as the sun's rays fell upon them. It was the picture of a desert, but if it be true that a picture is masterful in proportion to its power to stir the emotions, then the picture from that peak of the Panamints is not to be compared with any tawdry scene that needs the colors of vegetation to make it attractive.

Mr. Henderson, because of the vast space which the eye could cover there, named the mountain Telescope Peak.

After the visit of Mr. Henderson, Death Valley was abandoned to the Piutes and the renegade whites who lived with them. Very likely the Mormons may have crossed that way, now and then, en route from Utah to Southern

California, but the American citizens had all they could do to preserve their homes and settle, once for all, a trouble that nothing but blood would settle. And when the war was over, it took so long to adjust themselves to the old-fashioned prospecting again, that it was not until about the year 1870 that the hunt for the Gunsight lead was resumed with pristine vigor.

At last the time came when the prospector thought he had it, or the next thing to it, and in 1873 was organized the Panamint Mining District, while a roaring camp of thousands sprang into existence in what was called Surprise Cañon, on the west side of the Panamints.

There never was a cañon more appropriately named. The prospector was surprised when he found the lead, the mining sharps were surprised to hear of his luck, the growth of the camp was more surprising still, while the way the bottom dropped out later on fairly took the breath of everybody concerned. While the fun lasted, however, the mountains round about were pretty well tramped over and perhaps prospected. Many traversed Death Valley in all directions, and a few left their bones there because they were ignorant or foolhardy.

Then Death Valley had another rest from the rush of prospectors, though one "Cub" Lee, a white Arab with a Piute wife, and his brother Philander "held a bunch of cattle about Furnace Creek for a year," or more, while "Bellerin" Teck took a ditch full of water out of Furnace Creek and made a small ranch on which "he raised alfalfa, barley, and quails." "Bellerin" Teck got a reputation for being a bad man. He traded a part of his ranch to a Mormon named Jackson for a yoke of oxen, and then within a week ran the saint out of the valley with a shot-gun. Then "Bellerin" moved out himself and, sad to say, faded out of Death Valley history. A man who rejoiced

VIEW ON FURNACE CREEK.

in such a name, who was handy with a shotgun, and who withal was the first citizen of Death Valley, must have had an interesting biography.

Then there was the exploring trip of Lieut. Wheeler in 1871. No story of Death Valley would be complete without reference to the Lieutenant. While in the valley, in the heated term, if one may believe what "Cub" Lee says, he ordered his guide to go across the valley afoot, on some errand which the guide declared impossible. So the Lieutenant called two soldiers, who, with their bayonets, compelled the guide to start. Inside of two hours one of the soldiers staggered back into camp, just able to walk. A relief squad carried the other soldier, but the guide had become insane, wandered away, and never was found.

Last of all came the discovery of borax by Aaron and Rosie Winters, as told elsewhere. When the claim of Winters had been purchased, W. T. Coleman and F. M. Smith, being partners in some borax operations, started in to work the deposit north of Furnace Creek, while the Eagle Borax Company, of which J. Daunet, mentioned elsewhere, was president, began on the deposit near Bennett's well. Surveyors to stake out the claims of the borax people were sent to Death Valley, and then, for the first time, definite and accurate statements began to reach a limited part of the public. First of all, there was a statement about a previous survey supposed to have been made by the Government. Says Mr. J. J. McGillivray:

"The San Bernardino meridian line traverses the valley, running due north. It is supposed to have all been surveyed once, and the Government paid large sums to contractors for the work. I attempted to retrace the fifth standard parallel, which is supposed to have been surveyed, but found that the work had never actually been done. None of the townships had been staked out, and what is described on the Government maps as low and level land, is 8,000 feet high, and at an angle of forty-five degrees. There has always been more politics than straight lines in California surveys."

On getting their claims in shape, the borax manufacturers went to work to develop the deposit. Pans in which to boil the material, vats in which to crystallize the product, pipes to bring water from a spring, and lumber for houses, stables, etc., had all to be hauled out to the valley from San Bernardino, 250 miles away, over the desert, with never a house, and scarce a spring between. It cost 8 cents a pound to carry the stuff across the desert; moreover, men had to be secured to do the work in the valley. But to the Californian, such difficulties as these problems presented were matters of common occurrence, though it must be admitted that Death Valley was a little worse than any other proposition the coast miner had had presented to him. The works were erected, a mile and a half of good water-pipe put down, houses for men and animals were built, and the work of making borax went right along.

Nor was that all. The company imitated, or rather followed, the example of Mr. "Bellerin" Teck, Death Valley's first settler, and established a ranch, and named it Greenland. That was a descriptive name, in a way, certainly, though the ranch in no way resembles the frozen country to which the mind turns when the name is spoken. Perhaps had they been versed in Greenland lore, they might have called the ranch Ivigtut, for that is the Greenlander's term for a green vale in a barren region. Anyhow, a green spot was, and is, the ranch in Death Valley, with its half-acre pond, its thirty acres of alfalfa and trees, its 'dobe house with a wide veranda and its running water on all sides. For five years the ranch and the works were run, beginning in 1883. Then Coleman went broke and the works shut down, and only one man now lives in Death Valley—James Dayton—whom the Arabs call a sailor, because he was once cook on a Sacramento River steamer.

The story of the valley during those five years is quickly summed up, because it was devoid of special incident. On the average, forty men were employed; teams were coming and going all the time, and the tramp and other prospectors made the works, as far as allowed to do so, a center of operations. The men had the best food obtainable, and were housed as men are in all mining camps. On the whole, Death Valley was a pretty fair camp, considering the fact that "Piute ladies," as an Arab said to me, were the only women who visited the place. But the men could never be made contented, even in the winter season. It was healthy then, and the pay was good, but the miner, even the desert placer miner, is a gregarious animal, and Death Valley, from a social point of view, had its drawbacks. The man who had lived there six months was an old citizen, while one who remained from September, when the work for each season began, until June, when the heat compelled the closing of the works, was a marvel to the rest.

How thoroughly the region was prospected during this time can not now be told, but the Gunsight lead was never found. As a rule, mining men scout the idea of there ever having been a Gunsight lead. Silver is rarely found pure, or approximately so, and men who are flying for their lives out of a desert land do not weigh themselves down, even with rocks that seem to have metal in them. But others say that while the mine was found, it can not now, and probably never will, be located, because the spring and the cropping have been buried out of sight by some cloud-burst.

The topographical features of Death Valley, as will be

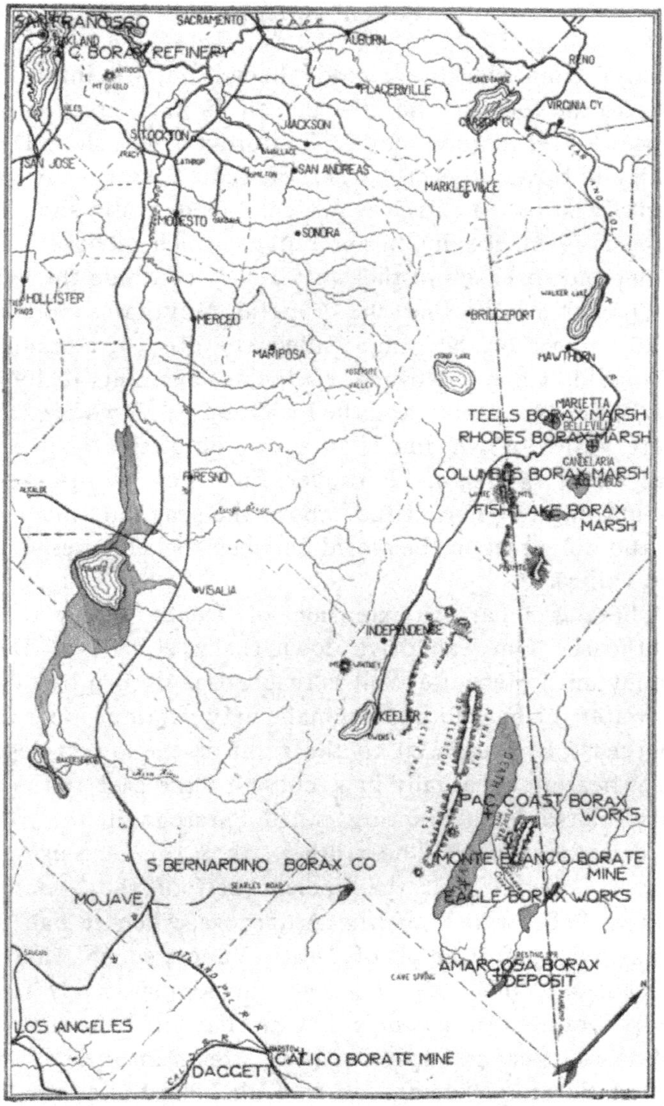

MAP OF THE BORAX FIELDS OF THE PACIFIC COAST.

inferred from what has already been said, are interesting even to the tourist, who, like myself, is not versed in geological lore. A good map of California would show Death Valley in the southeast corner of Inyo County. A good map of California and Nevada combined would show that extensions of the mountain ranges which inclose Death Valley on the east and the west, really continue the valley far into Nevada—that the Funeral Mountains are practically a part of the range called the Red Mountains, in Esmeralda County, Nevada, while the Panamints really join the White Mountains, and the fertile and well-watered Fish Lake Valley of Nevada is an extension of the terrible and arid Valley of Death. However, Fish Lake Valley lies at about a mile higher altitude above the sea, and this makes all the difference in the world between the climates of the two valleys.

There is a narrow extension of Death Valley to the southeast. One can drive down the west side of Death Valley until Mesquite Well is reached, and then by taking on water for himself and animals, may continue on to the southeast, keeping well to the right of the lowest depression, bearing eventually in a curve to the east and northeast, and so reach a spring called Saratoga, in the valley of the Amargosa. The truth is that this extension of Death Valley is but the ancient bed of the Amargosa River. The waters of the Amargosa, when it had any, used to flow into the pit of Death Valley, which was then a lake, and some say that even in comparatively recent years a rain-storm has been known that filled the river bed for a brief time and sent a roaring stream into the old sink. But at Saratoga Springs, as in Fish Lake Valley, the altitude is much higher than in Death Valley proper, and the climate, though bad enough to be dreaded by all, is not as bad as that of Death Valley.

Death Valley proper is unique. It is about seventy-five miles long, running from north to south, and from five to fifteen miles wide. At its lowest point, where its climate is worst, the width is not above eight miles from foot-hills to foot-hills. It is to the west of, and opposite to, this depression that the Panamints reach their highest altitude, while on the east, the Funeral Range is practically one huge ridge, with almost a vertical precipice on the side next to the valley. A few miles to the south, a mountain range running east and west shuts in the foot of the valley, so that at its lower end Death Valley is walled in on all sides but one.

Just what the depth at the lowest depression is, I do not know. A California mining bureau report, written by Prof. Henry G. Hanks, puts the lowest depression at 110 feet below the sea. One of Dr. C. Hart Merriam's party of Government experts, who went into the valley in the summer of 1891, said the depression was 200 feet below the sea. I have seen one statement in print which placed the depression 400 feet below the sea. No doubt that was an exaggeration. Whatever the real depression is, it is interesting to note, as Surveyor McGillivray pointed out, that fifteen miles to the west of this depression was Telescope Peak, rising two miles above the sea, while within an equal distance easterly, was a Funeral Peak rising 8,000 feet above the sea. Where can two such mountains like these be found, with such a rent as this between them?

Unique as is this little pit of desert in its depth and surroundings, the reader should be told, in case he ever visits it, that it is likely to disappoint him, because the eye does not realize how high the mountains are, nor does it see that the bottom of the pit is below the sea; nor do the mountains on either side look to one in the old trail to be so extremely precipitous as they are. One has to wander around, try to scale the mountain-faces, and compare what he sees with other mountain wonders, to fully comprehend Death Valley as a valley.

The foot-hills, wherever they occur on the east side of the valley, but especially for several miles north of the mouth of Furnace Creek, are such that no tourist could fail to give them attention. They are, for the most part, clay buttes and

MUSHROOM ROCK.

lava peaks, with here and there a butte covered with water-worn pebbles—shingle from some old-time beach. The wonderfully contrasted colors and the wind-worn, perhaps water-worn, forms of clay make one stop to gaze involuntarily. A natural curiosity, too, is Mushroom Rock, a singular block of lava, that has been worn, probably by sand, until it resembles the most common form to be found in cakes of ice floating off the coast of Greenland late in the summer—a form best pictured by putting a bunch of cauliflower on a mirror.

A very great part of the bed of the valley is formed of the pebbly wash from the inclosing mountains, but down its axis runs a salt marsh interesting to contemplate. In three or four places along the salt artery, where the higher land adjacent slopes down almost imperceptibly to it, wide fields of crude borax-borate of soda are formed in sandy crusts. But elsewhere the salt marsh is crusted over with the salt that is chloride of sodium. In most places this crust is very thin—neither man nor beast could cross it without snow-shoes. But in one place it is so thick and strong that a roadway was formed across it, making of it a bridge, of which a description is given in another chapter.

In its general aspect, Death Valley is gray and sombre; it is even desolate and forbidding. To admire the scenery from any point in the valley, one must have a taste for Nature in her sternest moods. The natural vegetation is scant and stunted, and there is not a green thing that grows there naturally. The thorny mesquite trees are of a yellowish-green tinge; so, too, are the grease-bushes, while the sage-brush and weeds, of which there are several varieties, are either yellowish-gray, or the color of ashes. A little round gourd grows in some of the cañons. It turns yellow, when ripe, and has a thin meat within that is exceedingly bitter. It is called the desert apple. The cactus that grows beyond the valley in abundance is rare here. In short, the vegetation of Death Valley is terribly scant in comparison with that of even the Mojave Desert.

Arid as the valley is throughout its whole extent, there are two running streams within its confines. One comes in at the north end, where it forms a marsh that gives out volumes of sulphureted hydrogen. Some who have seen

3

AN OASIS—DEATH VALLEY RANCH.

it believe that the water comes through a subterranean passage from Owens Lake, beyond the Panamint Mountains. The water of this stream is like that of the lake, and the flow never varies from one season to the other. Incredible as the proposition seems, this brook may be an outlet of the lake.

The other stream is Furnace Creek, mentioned elsewhere, which rises in springs in the Funeral Mountains, has pretty good—if warm—water, and is the only possible support of the ranch that was made by the borax people.

But more interesting to the tourist than all that can be seen or said about the lay of the land, are the stories told about its climate. As was said at the first, the story of Death Valley is full of apparent contradictions. Here was a ranch, for instance, on which three men found work in caring for the meadows and stock; a little over a mile away were the buildings where forty men were employed, most of them in the open air, wholly unprotected from the sun's rays, and some engaged about a furnace where a great heat was maintained. How could these things be if it were true that men died from heat and lack of moisture when they had water in their hands? It was a curious case, but both statements of facts were true.

With the prevailing wind from the west, Death Valley, deep and narrow, is guarded on the west by the lofty and precipitous Panamints, while four other ranges and four valleys, for the most part absolutely arid, lie between it and the sea, the only source of moisture. Even west of the Sierras, the plains of Tulare County must be irrigated to make them productive. Imagine now what the condition of the air must be, when having been drained of its moisture by the ranges near the sea, it sweeps inland over the wide and undulating desert east of the Sierras, where the sun's rays beat down relentlessly from above, and are reflected back up from yellow mesas and white hot salt beds! It becomes not only so hot that it strikes the face like a blast from a furnace; it is well-nigh devoid of moisture. People who talk to the weather sharps of the Signal Service Bureau, are told that with 90 per cent. or more of humidity in the air in summer, the weather is insufferably oppressive; with 70 per cent., the air is about right; with but 60 or 50 per cent., as when the air in a room is heated by a stove or furnace, the moisture is taken from the body in a way to produce headaches, but should the percentage be reduced to 40 or to 30, the air becomes positively dangerous to health. In Death Valley the air, raised to furnace heat by its passage over the deserts, is kiln-dried in the pit below sea level, till the percentage of moisture is at times said to be less than 1.

Of the effect of this heat, abundant and trustworthy testimony may be had.

While making the ditch which supplied the ranch with water, J. S. Crouch and O. Watkins slept in the running water, with their heads on stones to keep their faces above the fluid, although the work was not done in the hottest season. Philander Lee, an old desert Arab, well accustomed to the heat, while at work on the ranch, regularly slept in the alfalfa where it grew under the shade of some willows and was abundantly irrigated.

Other effects of the arid air are found in the utter ruin, within a few days, of every article of furniture built elsewhere and carried there. A writing desk curled and split and fell to pieces. Tables warped into curious shapes. Chairs fell apart. Water barrels, incautiously left empty, lost their hoops in an hour. One end of a blanket that had been washed, was found to have dried while the other end was manipulated in the tub. A handkerchief taken from the tub and held up to the sun, dried in a flash—quicker than it would have done before a red-hot stove.

Meat killed at night and cooked at 6 in the morning had spoiled at 9. Cut thin, dipped in hot brine and hung in the sun, it is cured in an hour. Flour breeds worms in less than a week. Eggs are roasted in the sand. Fig trees bloom and produce fruit near the house every spring, but the figs never mature. Though water flows about the roots of the trees, the figs dry up and fall off in July.

Surveyor McGillivray said, after running out the land for the borax companies:

"The heat there is intense. A man can not go an hour without water without becoming insane. While we were surveying there we had the same wooden-case thermometer that is used by the Signal Service. It was hung in the shade on the side of our shed, with the only stream in the country flowing directly under it, and it repeatedly registered 130°, and for forty-eight hours in 1883, whenI was surveying there, the thermometer never once went below 104°.

"Several of our men went insane. One of them was a Chinaman, who had wandered away as soon as he had lost his senses. We hunted for him awhile, and were then forced to give him up as lost. A few days afterward we went to a town sixty or seventy miles from there to get some provisions, when an Indian came into the town leading our lost Chinaman, still insane, and performing all sorts of strange tricks, to the infinite delight of the Indian, who thought he had found a prize clown, and regarded it as the best joke of the season."

The human body, when suffering from a fever, is dangerously hot at 105° Fahrenheit. It has been known to reach 112°, but death quickly followed. A thermometer hanging under the wide veranda on the north

BORAX WORKS, IN DEATH VALLEY.

side of the adobe house on the Death Valley ranch, has registered 137°. It is in such weather as this that the sand-storms in their deadly fury sweep through the valley, and even desert birds, caught away from the saving spring, fall down and die. It is a fact that since the ranch was established, one man has died from the heat while lying still in the house; and another, while riding, with a canteen in his hand, on top of a load of borax bound down the valley, fell over and expired. "He was that parched, his head cracked open over the top," said a man who saw the body.

Such is Death Valley in the heat of summer. No work worth mentioning was done there, or ever will be done there, at that time of the year.

On the other hand, Death Valley in October becomes a dreamy, sunny climate, the home of the Indian summer. The change of climate which the whole desert country undergoes in the course of a year is remarkable. One reads in the authentic reports of the California Mining Bureau about snow falling in the mountains west of Death Valley to a depth of three feet, while Superintendent Strachan, of the Teels Marsh Borax Works, in Esmeralda County, Nev., noted a temperature of 120° in the shade of his house in August, and yet, before the winter was over, saw mercury freeze and the temperature sink to 50° below zero. There is probably no place on earth where a wider variation of the thermometer than this has ever been observed, just as no place so hot as Death Valley has been found, the greatest recorded heat of the arid region about the Red Sea being less than 127°.

But one feature of Death Valley weather remains to be noted here, though much more of its history will be told in the following chapters. It will not do to say that rain never falls in Death Valley; it rarely falls there, but cloud-bursts—concentrated storms of the utmost fury—are often seen about the mountain tops, as well as around the mountains throughout the desert region. As described by the desert men, they come in the hottest weather, and usually when least expected. Right in the clear sky appears a cloud, black and ominous, streaked with fire, growing with wonderful rapidity, and eventually sagging down like a great sack. The cloud is always formed above the mountains, and after a time its bulbous, sagging body strikes a peak. Floods of water are released on the instant, and in waves of incredible size, they roll down the cliffs and cañons. Precipices and peaks are carried away, gulches are filled with the debris, mesas and foot-hills are covered. The face of a mountain may be so changed in an hour as to be scarcely recognizable, and even the lighter storms rip the heart out of a cañon so that only jagged gulches and heaps of broken rock are found where once, perhaps, a good trail existed.

"Cub" Lee tells of sleeping near the mouth of Furnace Creek Cañon one night, years ago, with "a bug hunter"(as the desert-tramping scientists are called) in camp. It was so hot that the bug hunter could not sleep. About midnight, he heard a roaring noise up the cañon, which, as it kept increasing in volume, caused him to look up that way. To his surprise he saw, as he supposed, the sky that appeared between the cañon walls grown suddenly white. At that moment Lee rolled over and the "bug hunter" asked him what ailed the sky. Lee gave one glance and then yelled:

"Cloud-burst, climb!"

They scrambled up the steep wall, as best they might, just in time to save their lives. Lee thinks the foaming wall of water that had whitened the sky was not less than 100 feet high.

CHAPTER II.

SAND-STORMS AND SAND-WAVES.

DURING the year 1891 an expert Signal officer, Mr. John H. Clery, was sent into Death Valley by the Government in order that he might make a note of and write a report on the meteorological conditions prevailing in that wonderful region. It is a pity that the observations of such an officer should be buried where the public can never see them —in a public document; but because they will be so buried from sight I may venture to tell here what I saw and was told about the sand-storms that rage there nearly, if not quite, half the time.

I awoke there one morning to find the wind blowing from the north down the valley at perhaps twenty-five miles an hour. The sun was shining dimly, and the house and trees at the ranch made vague and at times very indistinct shadows on the ground. To the east, particularly where the sun was shining above the peaks of the Funeral Mountains, the air was full of a whitish haze, such as one may see at sea when a storm is gathering. To the west the landscape was blotted out by a dense brown fog. Within a few miles west of the Furnace Creek ranch, where I camped, the peaks of the Panamint Mountains rose from 6,000 to 8,000 feet above the level of the valley, but neither foot-hill, nor face, nor snowy crest could be seen. Indeed, neither grease-bush nor mesquite tree could be seen in the valley half a mile away in that direction, nor was the vision much less obstructed to the north and south. A sand-storm was raging throughout the valley, and the air was smoked with the dust so high that even the highest mountains were obscured.

People throughout the East think they are familiar with dust. They see it picked up by the wind from country roads and city streets, and carried along by the blast in blinding clouds. They see the particles of dust, sometimes in puffs, sometimes in serried ranks, drive past the trees and fences and houses, leaving a film over all as they go. They see tiny tornadoes, the whirligigs that on the street corners pick up the debris and toss it aloft in a merry-go-round, that they will avoid if possible. But nowhere in the East can anyone see anything that may be compared with the sand-storms in Death Valley. One indeed may get a faint idea of what such a storm looks like from a distance, if he will stand on the deck of a lake vessel and gaze over fifteen or twenty miles of water at the clouds of smoke and dust that hangs above Buffalo and Chicago. Or he can go to the palisades at Fort Lee, on the Hudson, on a windy day, and find a faint resemblance in the dust that hangs over the metropolis.

As was said, when one looks at a dusty squall in the East, he sees the particles and debris driving with the wind past the objects which the wind can not move. In a sand-storm in Death Valley, one has to stare a long time into the murky atmosphere before he can distinguish any moving object. Even though the wind be blowing at great speed, it is as if one were trying to look through a Bay of Fundy fog on a windless day. It was thus that I first saw a Death Valley sand-storm. I could not see where the dust that filled the air came from; in fact, I should not at first sight have thought there was any dust in the air, had I not previously been told some of the characteristics of such a storm. I should have thought rather, that a curious brown fog prevailed. After a while, however, I noticed the dry taste and grit of dust which the breath drew into the mouth, and that was a pretty good proof of the character of the fog. It is true that little gusts of wind dashed along now and then, picking sand and even little pebbles from about the clump of bushes or mesquite trees where eddies were made in the wind, and little puffy clouds were formed there as they are when the wind picks dust from the eastern roads, but these clouds quickly fell back to the earth or were dissipated. The general or ordinary power of the gale seemed not to move anything but the branches of the shrubs.

Looking toward the east I observed that the white haze thickened and thinned over the face of the sun, as if clouds of varying density were passing there, though no distant clouds could be seen. The face of the sun was a great white disk, that faded and brightened alternately without entirely vanishing or becoming too bright to look at, and yet no outline of a cloud could be distinguished. That was curious, but more curious and interesting was the view when looking toward the sky above the center of the valley. It was along the axis of the valley that the dust fog was densest, and here its highest strata seemed frayed out into waving feathery streamers, while the dark brown masses below them showed light and dark shades of density. What with the sun showing ghostly white, and a seemingly stationary fog-bank in a gale of wind, the spectacle was singularly strange and weird.

But if strange from within, what shall be said of the sand-storm as seen from the top of one of the higher mountains that border the valley? The whole vast space between the ranges was filled with smoky billows, ragged and torn and tossed about, and rising up till all the foot-hills and half the heights themselves were covered. Sharp squalls plunged down the cañons and gulches, and there gathering the dusty forms in their arms, went whirling away in gigantic waltzes. It is no wonder that the Arabs of this desert country, the Piutes, believe in witches and supernatural powers in the air.

It was these irregular whirling gusts of wind that formed the dusty fog into the feathery banners seen from below, and it was they, too, very likely, that gathered the great mass of the dust up from plain and mountain side to fill the air.

At times these lofty whirligigs take on more substantial form, become in fact like the water-spouts at sea—become sand-spouts. The desert men there call them sand-augers. It is a marvelous spectacle when a sand-auger travels down the valley, particularly if its course be along one side of the valley where the dusty fog is not so dense as elsewhere. Astonishingly slender in form, it rises writhing and twisting, sometimes a mile in the air. With a faint puff or cloud at the top and a slight spread at the base, away it goes, sagging and swaying hither and thither, its sinuous grace fascinating the eye as it travels, until somehow it all unexpectedly vanishes out of sight. I saw two such augers. Both were more than 2,000 feet high. Neither journeyed a mile, but when they faded away there was no sign of a falling cloud or a thickening of the dust in the air where they had been. They simply disappeared ghost-fashion, as I was looking at them.

Curiously, as it seemed to me, the ordinary whirls in the air-whirls, too, that feathered out the mass of the dust fog—could not be seen when in the valley, save as one saw the streamers. At least, during a drive of about five miles through the storm, I failed to observe any indication of their existence, save now and then a sudden shift in the direction of the wind, which lasted for a brief interval only.

The sand-storms commonly last for three days. The one I saw lasted but two. In the afternoon of the first day the speed of the wind increased rapidly, until it was probably traveling at times as high as fifty miles an hour—perhaps more. At this time the gusts picked up pebbles, as well as sand and dust. The air near the earth was full of flying missiles. No man could face it, and the horses grazing on an alfalfa field at the ranch, stopped feeding to huddle in bunches, tail to the wind and with heads down. It was like a blizzard of Dakota, with the grit of broken volcanic rock in the place of snow.

The reader has often observed that in certain conditions of the air the rays of light spread out from the sun, as it sinks in the western sky, like the leaves of a fan. Children on shore say that the sun is then drawing water; sailors at sea say there are back-stays to the sun. The devil may not be able to make a rope of sand, but the power behind a Death Valley storm makes back-stays to the sun out of dust. As the sun went down behind the Panamint Mountains on those stormy days, the long white rays shot out in all directions, making a sun-burst that was strikingly distinct, and as beautiful as it was ominous.

It was a winter storm that I saw, and certainly one of but moderate strength, and yet, as I say, though the weather was not very cold, the storm could properly be compared with a Dakota blizzard in its effect on animals exposed to its fury. It was really a terrible storm, but with his hands gloved and his face well wrapped up, and good goggles to protect his eyes, a man, because the temperature was somewhat above the freezing point, could have walked across the valley in the worst of it. But in August the case is different. Day after day the sun beats down from a cloudless sky on this desert pit. The earth grows hotter and hotter as the days pass, till the naked hand can not bear to touch even the woody parts of the sage and grease-bushes, while a rock or a bit of iron scorches as if from a bed of coals. The air glows, and dances, and bakes. And then comes the cloudless gale, gathering heat as it hurries along between the blazing sun and the white, hot sands of the desert, until at last it bursts in through the passes of the Panamints like a blast of flame. There it picks up the hot dust and sand, and tossing them aloft in clouds, converts the valley into a veritable pit of hell.

As was said in the last chapter, people have read, from time to time, that men and beasts, and even birds, trying to cross Death Valley, fall down and die. It is true. Even the desert linnet, whose home is on these arid wastes, sometimes succumbs to the terrors amid which it lives. When caught in the midst of a Death Valley sand-storm in the heat of the summer, these birds fall gasping from the mesquite trees, in which they vainly seek shelter, and die. Life is burned out of them. It is not that the air is poisonous; the air is simply dried and heated till no living organism higher than that of a reptile can endure it.

I did not see this, of course, but Mr. J. W. S. Perry, the superintendent of the Pacific Borax Company's Works, at Daggett, a man of education, and of experience as well, and Leander Lee, an old desert man whose home is in the Amargosa Valley, both told me that they had seen birds dead and dying under less trying circumstances than those which prevail in an August sand-storm, and I believed them. Moreover, it is on record in a report written by Henry G. Hanks, as State Mineralogist of California, that Mr. R. R. Hawkins, who visited Death Valley in 1882, saw the same phenomenon, while two prospectors who tried to cross the valley in the same year, fell down and died, although they had plenty of water with them. The terrors of Death Valley have been misunderstood and grossly misrepresented; but, considered as they are when at their worst—that is when the summer sand-storm is raging—they have never been exaggerated. It is simply impossible to adequately portray such a storm, let alone exaggerate in describing it.

As I said, a sand-storm, more or less severe, is making Death Valley uncomfortable about half the time. I have this statement from several men familiar with Death Valley and the desert region round about. What shall be said then, of the following quotation from the report of the State Mineralogist just referred to:

"October 1, 1860, Dr. S. G. George, Dr. W. B. Lilley, T. J. Henderson, Stephen Gregg, Mr. Thayer, and J. R. Bill, organized a search for the Gunsight lead. They * * * * remained at the emigrant camp for some time, prospecting the hills in every direction. Although ten years had passed, the tracks of men, women and children were distinctly seen, as fresh as if newly made; the irons of the wagons were where they had been left. The remains of ox-yokes were seen, which had been laid out for use on the following day, with the chains extended on the ground in front of each wagon, showing the number of oxen to each, and traces of the old camp-fires were seen."

With sand-storms raging half the time so that the air is full of drifting dust, with sand-augers—slender tornadoes, in fact—drifting about sucking up the sand; with a "burning wind, fierce and powerful, blowing articles of considerable weight some distance, and hurling the coarse hot sand with such force as to lacerate the face when exposed, the men being frequently obliged to wear veils and goggles," as the report elsewhere says, how did it happen that "although ten years had passed, the tracks of men, women, and children were distinctly seen?"

The facts alleged can not be doubted by any one who

visits Death Valley. I did not see the tracks referred to nor very much of the debris of the old camp. The tracks had been obliterated and the remains of the camp carried off by the men employed in the valley (during the cooler months) when borax was produced there, and by the prospectors attracted there by the presence of a civilized gang of men with supplies. But I did see a sand-storm and sand-augers, and I saw the old wagon trails made years ago and unused for years, but still as distinct as though but a month old. Moreover, I saw sand dunes, though comparatively small ones, wherever I saw a growth of mesquite trees, and they were dunes that grew fast enough and large enough to cover big clumps of these trees. Why does not the sand obliterate tracks and trails?

The answer as given by the desert men is that the sand there is of such a nature that when impressed by the wheel of a vehicle, or even trodden by the foot of a child, the particles are compressed, not to say felted, together, so that the wind has little, if any, effect upon it. The wind does, indeed they say, blow sand and dust into trails and tracks, but it blows them out again and leaves the mark undisturbed. But when the sand is blown among the mesquite trees, the branches hold the accumulation and so are buried.

But if the storms of Death Valley fail to pile up any great quantity of sand, in the shape of dunes, the winds in a part of the Amargosa Valley to the east make up for the lack. I have had the good fortune to see some remarkable heaps of sand in various parts of the United States. There is such a heap, a veritable wave of sand, fifty feet high and a mile and a half long, at Cape Henlopen, that is rolling back from the sea and burying a forest as it goes. There is another sand-wave on the long slender island north of Cape Hatteras, between the Pamlico Sound and the Atlantic Ocean. This sand-wave, though no more than fifteen feet high, is not only destroying all the trees and every vestige of vegetation in its pathway, but it is driving from their humble homes the entire population, some hundreds of souls, of that part of the island. More remarkable in themselves (though less in surrounding conditions, because they are destroying nothing of consequence) are the sand-waves in the Snake River Valley in Eastern Idaho. The tourist may find on the north side of the north fork of the river, a series of sand-waves thirty miles long, that are for all the world like rollers on a shelving seacoast. They vary in height from twenty-five to more than three hundred feet. The prevailing wind is from the southwest; it blows up the valley. And these waves are traveling up the valley before it. It is a curious fact that these waves are composed of fine white sand, the like of which is not found elsewhere in that vicinity, and they are traveling over a continuous bed of black lava.

Quite as interesting, though much less in extent, are the sand-waves in Amargosa Valley, near the Saratoga Spring. They are seen, white and beautiful, by the tourist as he travels down the cañon on the usual route, long before he reaches the valley. They lie on the further or northerly side of the valley, and because of their shape and the immense masses of black lava under and beyond them, they at once suggest those to be seen in the Snake River Valley.

But the most striking feature of this panorama of sand-waves is seen off to the left as the traveler descends the cañon. Jutting out from the mountains on the further side of the valley, out directly across the path of the sand-waves, is a row of black lava peaks from five hundred to seven hundred feet high. It is as if these peaks had suddenly been thrown up there as a barrier to check the sandy tide, but the sand-waves have simply rolled up on the weather side of the barrier, up and up, until they have broken over the highest peaks and have tumbled, white as foam, down on the lee side of this black reef.

I have seen the dark waves of the Atlantic breaking against the formidable cliffs of Cape Desolation, on the coast of Greenland, and I have seen the sunny rollers from over the South Sea crash in foaming masses on the dreamy shores of the Bay of San Juan del Sur in Nicaragua, but the one was not more impressive, nor the other more beautiful than this smother and froth of sand on the black lava reef of the Amargosa.

CHAPTER III.

ONCE A CITIZEN OF DEATH VALLEY.

A BRIEF chapter in the story of Death Valley should be devoted to an unfortunate Frenchman named Isidore Daunet. Daunet was born in Bassis, Pyrenus, France, on April 4, 1850. He emigrated to California when ten years old, finding a home in San Francisco until 1863. Then, although but a boy, he began to wander up and down the mining camp region, picking up a knowledge of ores and eventually becoming, while yet not of age, a typical prospector. When fully grown he was noted as a remarkable specimen of manly strength and vigor, and was moreover possessed of a great courage and energy of character, qualities which saved his life in a trip through Death Valley when others died, though they failed him nevertheless at a critical period later on.

In the year 1880, Daunet was in the mining camp of Panamint, on the west side of the Panamint Range, and but a few miles in a direct line from the most depressed portion of Death Valley. Finding no prospect of striking a lead there that would make him rich, he joined a party of other adventurous spirits bound on a prospecting trip into the deserts of Arizona. The party numbered seven men in all, and after buying supplies and getting pack animals together, they started away, although it was in mid-summer, by an unfamiliar trail that next led them into Death Valley.

As they went down the cañon they knew very well that

it was Death Valley that lay before them, but they were in the prime of life and health, and scouted the idea that they could not pass across its narrow breath. Nevertheless as its arid atmosphere sapped the moisture from their bodies they strove in vain to supply the lack by drinking from their canteens. Almost before they realized their condition their water was gone, they had no knowledge of the location of the springs there, and half wild through their suffering, they cut the throats of their pack animals and drank the spurting blood, as tigers might have done.

Then Daunet and another, the strongest and most resolute men in the party, started off for help, and after shocking hardships, reached an Indian camp where water was abundant. The Indians at once returned to help those left behind, but found only two of them living; the other three had perished for want of water.

Not long after this incident, the story of the borax find of old Aaron Winters, in the upper end of Death Valley, was told throughout the region, and Daunet heard it. He had had a hard experience in Death Valley, but associating with himself J. M. McDonald, M. Harmon, and C. C. Blanch, he went down near the lowest part of the valley and secured 260 acres of good borax land. There was an abundance of mesquite wood for fuel, and water was got by digging. They carried in a boiling pan and crystallizing tanks, suitable for preparing borax for market, and by the end of 1882 had turned out 260,000 pounds, for which they received 10 cents a pound and upwards. Their first shipment was thirty-seven tons of crude material, and for that they got 8 cents. Nevertheless the enterprise being so far away on the desert from Daggett, the nearest point of shipment, they could not succeed in competition with more favorably located concerns.

On October 1, 1882, when at the height of what seemed

EAGLE BORAX DEPOSIT IN DEATH VALLEY.

a very prosperous career, Daunet was married to Clotilde Garraul, a French Canadian woman. Clotilde had made one matrimonial adventure already, although Daunet did not know it, but she had been divorced. Everything went on pretty well with the couple until 1884, when business troubles accumulated, and then there were quarrels in the Daunet family. In May these troubles culminated in the wife leaving Daunet and applying for a divorce. The service of the papers on Daunet was made on his arrival in San Francisco from his works in Death Valley. He had hoped to effect a reconciliation with his wife, and the shock to his nerves when the papers were served was too much for him. Going to his lodgings at 535 Post Street, on the morning of the 28th, he wrote a rambling letter "To The Public," tied up his head with a white handkerchief, sat down facing the mirror and fired a ball through his brain.

The borax plant and grounds were eventually sold and M. Harmon, one of Daunet's partners, is now (1892) running a restaurant in Daggett.

CHAPTER IV.

TALES OF THE WHITE ARABS.

NOTWITHSTANDING the fact that prospecting parties, some of them composed of educated, practical men, lured by the story of the Gunsight lead, visited Death Valley frequently after the fatal journey of the emigrant party in 1850, it was not until about thirty years had passed that any substance of commercial value was found within the limits of the valley, and then it was a citizen of the country, a genuine white Arab of this great American desert, who made the discovery, and the subsequent discoveries in the Furnace Creek cañon and the Amargosa were all made by men of the same class.

These citizens of the American desert are a remarkable class of frontiersmen in more respects than one, but in no respect more remarkable than in their choice of a home-site. Some of them have been there from twenty to thirty years—white men had been in Death Valley before the emigrant party that gave it its name perished there. Where they came from and why they went there, are questions not to be answered. It is not polite in desert society to ask questions of that kind. This rule, one may say, is *derigueur.* Society leaders there have been known to resent an infraction of desert social usages with a Winchester, but I apprehend that some of these men went there during the Civil War in an effort to escape the draft, some were deserters from the army, some went there because there were sheriffs with warrants in some other places, and some for the same reason that the old sailors drank whisky—because they liked it.

I guess Aaron Winters went there for two of these reasons. He certainly liked the country, and it is said that he had killed two men in his time. He has certainly killed one since, and how he did it is worth telling. Mr. Aaron Winters, it may be said, was a highly respected member of the most exclusive social circle of the desert.

In no way can I so well describe the class of men of which he is a type as by relating the stories I was told about him and the four or five who discovered the borax deposits in and about Death Valley.

In the year 1880, Winters was living with his wife Rosie in a valley known as Ash Meadows, just east of Death Valley. The name of the valley came from some stunted ash brush that once grew there. It was habitable for a family or two, because a little bunch-grass grew there on which a few cattle could feed; there were mesquite trees within twenty-five or thirty miles sufficient to supply an abundance of mesquite beans, which serve the Arabs in place of flour, and, more important than all the rest, there was a flowing spring of good water. Mr. C. M. Plumb, who visited it at the time, has preserved the following description of this odd frontier home of Aaron Winters:

"Close against the hill, one side half-hewn out of the rock, stood a low stone building, with a tule-thatched roof. The single room within was about fifteen feet square. In front was a canvas-covered addition of about the same size. The earth, somewhat cleared of broken rock originally there, served as a floor for both rooms. There was a door to the stone structure, and directly opposite this was a fire-place, while a cook-stove stood on a projecting rock at one side of it. At the right was a bed, and at the foot of the bed a few shelves for dishes. A cotton curtain was stretched over some clothing hanging on wooden pegs in the corner.

"On the other side was the lady's boudoir—a curiosity in its way. There was a window with a deep ledge there. A newspaper with a towel covered the ledge, in the center of which was a starch box supporting a small looking-glass. On each side of the mirror hung old brushes, badly worn bits of ribbon and some other fixings for the hair. Handy by was a lamp-mat, lying on another box, and covered with bottles of Hogan's Magnolia Balm, Felton's Gossamer for the Complexion, and Florida Water—all, alas, empty, but still cherished by the wife, a comely, delicate Spanish-American woman with frail health and little fitted for the privations of the desert.

"The shelves about the room and the rude mantel over the fire-place were spread with covers made of notched sheets of newspaper. Two rocking chairs had little tidies on their backs. The low flat pillows were covered with pillow shams and the bed itself with a tawny spread. In place of a library there were a number of copies of the *Police Gazette*. There was a flour barrel against the wall, a small bag of rice near by, and two or three sacks of horse feed in a corner. The sugar, coffee, and tea were kept under the bed.

"The water of the spring ran down the hill and formed a pool in front of the house, and here a number of ducks and chickens, with a pig and a big dog, formed a happy group, a group that rambled about in the house as well as romped beside the water of the spring. A few cattle grazed on the bunch-grass of the valley that stretched away before the house, gray and desolate."

It was just 200 miles across the desert from this home to the nearest settlement or railroad station.

One day, about the year 1880, a strolling prospector—one of the desert tramps—came along, bound, probably, from some Nevada town to Resting Springs, to eat up a grub-stake. He tarried over night at the Winters home, and told Winters a long story about the borax deposits up in Nevada, and what a great fortune awaited the man who could find more borax deposits. Winters was a shrewd fellow, and he asked many questions in a casual way and said nothing in return. Among other things, the prospector told him that one could test a supposed deposit of borax by pouring certain chemicals over some of the stuff and then firing the mixture. If it was borax the chemicals would burn with a green flame. Telling that was the only good thing that a tramp prospector ever did, so far as I learned.

When his guest had gone, Winters made haste to get chemicals. He had been in Death Valley more than once, had seen stuff there that answered the description of Nevada borax, and he was going to see what the Death Valley marsh held.

He took his wife with him, not only when he went after his chemicals but when he went prospecting in Death Valley. That was due to one of his peculiar characteristics. It happens sometimes that a long spell of rainy weather prevails over the desert in the spring of the year. When the rain at last clears off and the warm sun comes out, countless millions of plants spring up from the dust and sand, and the arid waste becomes one vast carpet of fragrant flowers. Aaron Winters was like the desert he lived on. His character was an arid waste in most respects, but he loved his wife.

Going over to Death Valley, this strange couple camped on Furnace Creek, and going down into the marsh gathered a small quantity of the most likely-looking deposit they could find, Winters "talking all the while and teetering and wabbling about," as was his habit when excited. Then they went back to camp and got supper, for the fire test could not be made by daylight.

At last the sun went down and the flaming colors in the western sky faded and darkened until the shadows in the gorge of the Funeral Mountains where Winters was camped became absolutely black. By the faint glow of a few dying coals Winters and his wife sat down on the sand, put a saucer of the material on a rock between them, poured the chemicals and alcohol over it, and then Winters scratched a match to fire the mixture. How would it burn? For years they had lived as the Piutes live on the desert. Not only had the wife to do without the little luxuries and comforts dear to a woman's heart; they had both lived on mesquite beans and chahwallas when the flour and bacon were gone—they had even gone hungry for lack of either. Would the match change all that? Winters held the blaze to the mixture in the saucer with a trembling hand and then shouted at the top of his voice: "She burns green, Rosie! We're rich, by ——."

They had found borax. William T. Coleman, noted as the leader of the San Francisco Vigilance Committee, and in other ways as well, was then a borax magnate. So was Mr. F. M. Smith. Soon after the news that it had burned green had reached San Francisco, two agents were sent by the firm of Coleman & Smith to the rude home in Ash Meadows. They found Winters a tough-fibered man, short in stature, stout in frame, dark-haired and with a full, florid face—past sixty years of age, but well preserved—in fact in every way a rugged frontiersman. He was slow of speech, somewhat reserved and unapproachable in manner, but a hearty, square man, bluff, brave, and generous. When it was understood that the new comers were there for business, Rosie got a bag of pine nuts somewhere in the camp, and while cracking and eating these around the camp fire the bargain was made. The deposit brought $20,000.

On getting his money, Winters went over to Pahrump oasis, in Nevada, and bought out one Charles Bennett, who had made a ranch there, bargaining to pay $20,000 for the outfit, of which $15,000 was cash in hand, the balance being covered by a fatal mortgage. Then he and Rosie sat down there and enjoyed life for a time, but the hardships previously endured had been too great for the wife. Prosperity came too late, and within two or three years she died.

One more characteristic story is related of Winters. It happened in the usual course that he had to go to Belmont, the county seat, one fall, on business—among other things to pay his taxes. It was a journey of several hundred miles, and Winters rather expected that some one would "hold him up" for what money he had along, and prepared for it by putting a worthless pistol in a holster on the dash-board of his buck-board, and a first-class Navy revolver under the cushion.

Sure enough, at a convenient place, as he neared Belmont, two men "got the drop on him," and he was obliged to get off the vehicle and deliver up his cash. This he did with much talk and palaver. He was going to Belmont to pay taxes, and it was all the money he had and all he could raise. If he didn't pay the taxes he'd be ruined, and wouldn't the gentlemen be kind to an old man and give it back. As he talked, he was "wabbling and teetering about" beside the buggy in his most nervous fashion. It made the road agents laugh to see him, made them laugh so that after a little they were thrown off their guard. Then one of them saw the worthless pistol on the dash-board and pulling it from the holster, turned and with a louder jeer than ever showed it to his partner.

At that, Winter's turn had come. In an instant he had drawn the revolver from under the cushion and shot one man dead, while the other, with his laugh turned into a chatter of fear, threw up his hands and begged for mercy.

Thereat, Winters disarmed him, made him put the corpse on the buck-board and then walk under the muzzle of the revolver into town. There the story was told and the robber, through the influence of Winters, was released, taken home to Pahrump and employed for more than a year as a ranch hand.

This was after the death of Rosie. They say that after Rosie died, Winters lost about all of his investment in the ranch.

The discovery of the Amargosa deposit followed naturally on that in Death Valley. Winters had a hand in this, but shared the good luck with two men named Parks and Ellis. Each of them got claims there, and the three sold out for $5,000. Parks took his money and went to his home in the East. Ellis died with his boots on at Pahrump the next year, 1883. It was in a characteristic desert row. Winters was then running the Pahrump ranch, and had a liquor store that was a resort for various kinds of citizens—white men, squaws, and bucks—who came sometimes a good deal more than 100 miles to have a spree. Ellis had previously killed a Spaniard in a mining camp, and in consequence carried a gun and a reputation as a bad man. One day, James Center, who had been a cook at the works then but recently established in Death Valley, went over to Pahrump to spend his accumulated wages, and there got into a game of poker in which Ellis had a hand. A quarrel over a jack-pot of just one dollar followed. Because Center accused him of cheating, Ellis went after a revolver.

He was intending to avenge the insult by killing Center, but Center grabbed a Henry rifle from behind the bar and went out to seek Ellis. Ellis was by this time returning, revolver in hand, and took shelter by crouching behind the wheels of a buck-board as soon as Center appeared. There he got the first shot, but he hit his man in the fleshy part of the right leg only, and Center was able to return the fire on the instant. His first shot pierced Ellis in the abdomen, through and through, and tumbled him over unconscious. He died two days later. Center was carried over to the Amargosa borax works a few days later, where Supt. Perry drew a silk handkerchief through the wound, dragged out the debris of drawers and dirt left there by the ball, and then cured him up. No arrest was ever made.

What is known as the Monte Blanco deposit of borates in the Furnace Creek cañon, was located by Philander Lee, Harry Spiller, and Billy Yount. Tradition does not make them noted in any other way. They never killed anybody, never got killed, never got wounded even. They just lived, found a deposit of borates and sold it for $4,000. Philander used his share in making a ranch at Resting Springs. The other two "went off somewheres." But Philander is an interesting fellow, for he has a squaw and several half-breeds, and a brother named Leander, who has another squaw with half-breed progeny, and two other brothers named Meander and Salamander, who have no regular squaws, but just live around among the Piutes.

So far as I could learn, no preacher has ever been among the Arabs of this desert, nor has any one been nearer to them than the mining camp of Candelaria, Nev., up in the northern part of it. When I asked about him one of the miners said:

"He stayed here three days—came Saturday morning—went away Monday night. We enjoyed it—yes, *sir*. He was a little red-headed youngster, and he went around and asked all the boys to come and hear him at the big dance hall at 7.30 P. M., *sharp*. We said we'd go and we went, though most of us got there a little late. When I got there he was rasping us for not keeping our promise to come sharp. He 'lowed we hadn't insulted him, but had insulted God.

"When I heard that, *I* thought we had better even up at once—had better repair the omission and insult *him* just mildly like. So the word was passed to chaw, and every man took his quid and the girls took gum. You see I'd heard that it made him sick to see a man chaw like his soul was in it, and I guess the proposition was a good one. You could see it working his countenance as he watched our jaws, and pretty soon he said he didn't feel well, and he was much obliged to the choir and we'd rise and be dismissed.

"Next day he felt better and was game to try a new deal. His proposition was to corner us one at a time and labor for our souls. We saw his pitch directly, and from that time until night he didn't get to say a word to a man—not one. Every soul kept an eye on him, and as soon as he'd start toward a bunch they'd slide. The bartenders dropped their glasses and things, the merchants their goods and everybody dropped everything and went out the back door and came out through an alley, and stood around looking as if nothing had happened. Why *we* enjoyed that parson's visit more'n a circus, but there never has been one here since."

CHAPTER V.

TRAMPS OF THE DESERT.

IN spite of its arid wastes, in spite of the discomforts, and in spite of the positive dangers to which even well-equipped travelers are exposed, the great desert region, of which Death Valley is the most noted portion, is not without its tramps. Human habitations are scattered scores of miles apart, and the country produces no food save such as a Piute Indian could live upon, but the lazy, thieving beggar finds his way up and down the length and breadth of the desert, refusing to work, yet somehow managing to live without it. The desert has even originated a species of tramp of its own, and it is an interesting as well as a novel branch of the worthless tribe.

As will appear elsewhere, I arrived at the old borax works, in the Amargosa Valley, after a day's drive of fifty-one miles, the greater part of which was through a most disagreeable storm of mixed rain and hail and snow. A more wretched day and night I have rarely experienced. The next morning the sun came out, but the air was, nevertheless, so cold that I could not walk about the buildings without an overcoat.

While taking such a walk, however, I saw the figure of a man coming along the trail that runs across the valley toward Resting Springs, a little oasis where there is an old disused quartz mill. The man was walking slowly, and, as he approached, I could see that he was limping.

His boots were run over at the heel in a fashion that would have made walking well nigh impossible for an ordinary man, but this one was, in addition, really lame in one foot. The boots were cracked open, showing stockingless and sore feet. His trousers were faded and frayed, his coat was full of holes, and his shirt plainly one which a miner somewhere had thrown away. In fact, his whole suit was made up of cast-off garments.

Over his shoulder he carried a bundle in which the forms of two or three cans could be seen, and in one hand he had an empty beer bottle. Reaching the old crystallizing vats at the borax works, he took a drink from the flowing well, and then going around to the sunny side of the vats, curled down on the sand and went to sleep. And there he lay from about 9 o'clock in the morning until 4.30 in the afternoon, without ever turning over or making a movement, save when a blast of wind would eddy around the vats and make him tremble and shiver.

At about 4.30 he roused up, and after painfully getting on his feet, rubbed his joints gingerly and then sat down again and took some scraps of food from his bundle and ate them. His meal did not occupy many minutes, and he soon went to a boy at the kitchen door and asked for some bread, saying he had been at work at a mining camp called Montgomery. He did not get anything, for the family living there had learned that it is very bad policy to favor a desert tramp. So filling his beer bottle with water, he started off over the trail toward Daggett, shivering with cold as he went. It was a trail 105 miles long, with not a house and but three springs in all its length. It was a trail he had never traveled; it was dim in places, and because the weather was freezing cold he was obliged to walk at night and sleep by day. I do not know whether he ever reached Daggett or not; I could learn nothing about him on my return there; but I suppose he pulled through unless he lost his way.

He was but a sample of a wretched host. They are not numerous by count, but in proportion to the widely scattered population they are more numerous than tramps are in New Jersey or Connecticut. Leander Lee, the old desert man who had charge of the works, drove with me over to Death Valley, but before leaving home he took his family to a brother's house at Resting Springs. It would have been unsafe to leave a woman alone there. Tramps were likely to come along at any time, and the desert tramp is more vicious, if possible, than the tramp of civilized districts. As the desert men did not have much to say when asked about instances of tramp brutality and the vengeance taken on the miscreants, I inferred that a skeleton might be found here and there among the rocks where a tramp had died through other causes than climatic effects.

But much more interesting than the foot-sore beggar limping along the trail, is the desert tramp that rides a burro—the tramp that has been evolved by the peculiar circumstances of mining-camp life—the grub-stake eating tramp prospector. Has the reader ever heard of the Breyfogle butte of gold, the Gunsight Lead, or the Peg-leg Mine? These myths are the stock-in-trade of the desert grub-stake eater. The stories are much alike. Emigrants en route to California became bewildered in the great desert. Their water gave out, their food was thrown away, their outfits were abandoned. In desperation they separated, and after untold hardships reached civilization worn to mere shadows of men. But each one had somewhere about his clothes a chunk of ore picked from a ledge as he wandered half in delirium, and that chunk of ore was literally loaded with free-milling gold, or it was a nugget of gold or a piece of native silver. The tramp prospector knows the stories by heart, and what is more, he knows within twenty miles of where that very ledge, or some other, is located. Now if somebody will only grub-stake him—supply him with the burros and food for say sixty days—he is just as sure to locate that ledge as the sun is to rise. He sits around the stores and saloons in mining-camps and tells his tale to all who will listen, and tells it with a most charming air of candor. He even goes to cities and tells it. He has dates and distances to fortify his assertions, and he refers to this and that lead he has, he says, discovered in other years. Are his victims the fresh arrivals in the camp—the tenderfoot from the East? Not at all. The merchant and his clerk, and even the mine-owner, yields to his seductions.

It is a matter easily understood if the reader has ever been in a mining-camp. There is no such place of possibilities in the world as that. Just to illustrate, let it be told that the little mining-camp of Calico, in San Bernardino County, California, had been in existence for years, and stacks of bullion had been taken from a mine there, when one day a stranger came to town and located a new mine right in the heart of town—a mine over which hundreds of mining experts had blindly tramped.

The plausible tale of the grub-staker sooner or later reaches a venturesome soul who will "go him once for luck." It is not a great risk—three burros cost say $50, and grub—bacon, flour, and beans for sixty days, with the rest of the outfit, less than $50 more. "What's a hundred, any how? It's like betting a bit against a twenty-dollar piece." May be this prospector really does know what he is talking about—he may strike something, anyhow, and a strike—here the imagination wanders away over the boundless delights that would follow should the prospector strike it rich.

The bargain is made. The prospector is very strenuous about the share of the stock he is to have in the mines that he finds. Not infrequently there is a wrangle, which but serves, as the prospector very well knows, to assure the capitalist of the sincerity of the prospector, but in course of time the outfit is complete and away he goes. He wanders out around the Calico range, he tarries at the Garlic Springs, he camps at the Caves, he loiters at Saratoga, he ambles past Amargosa, and brings up at Resting Springs. The name of this little oasis was derived from the favor with which the weary grub-stake eater looked upon it. It has for years been his Mecca. He has taken ten days to reach it, has tarried there forty-five days and has reached the settlements once more. He and his burros have grown fat, if the weather has been propitious, and they have brought sundry pieces of ore. Will the putter-up of grub-stakes look at the ore? He will. They are choice specimens. He will and does have them assayed. They show from 20 ounces to 1,500 ounces per ton, according to the character of the putter-up of grub-stakes. Some like what is called a low-grade proposition—a mine where there is a vast body of low-grade ore; others are caught by the hope of a mine of black metal that assays high. The prospector has sized up his man in advance and has had a sample all the time that would fit the victim. These samples are very easily obtained in any mining camp, the grub-stake victim knows it, and yet the prospector fools him with a tale of croppings and ledges in a new and most unexpected place. Of course they were found just as grub was giving out. Now if he could have sixty days to work on this find and develop it, the ore that he would get on the dump would be worth much more than the cost, not to mention the fact that it would be in condition to market or to show to capitalists who would furnish coin for a mill.

Lured on by the sample, the victim puts up grub for sixty days more—sometimes on one pretext and another for ninety days or for six months more. One young clerk in Daggett was caught for over $700 by a couple of wily desert tramps. In his case they salted a hole—there was but little ore required—and he made an inspection of it.

This is not the worst to be said of the grub-stake eater. He not only takes grub-stakes without making any return, but he willfully destroys property without reason or incentive. When the company owning the Death Valley borax deposits was operating its works there and running a line of great freight wagons over the road to Mojave, it had, as told elsewhere, a lot of stations or camps along the route at which the teams could get feed and water. While that route was open the prospectors bound for the Death Valley region were as thick as flies in a meat market. That they should feed and water their burros and make mush of the barley for their own use at the stations, was a matter of small moment, though in the aggregate they consumed many bushels every year. But more than a score of times they cleaned out the feed boxes, carrying off both barley and hay to some spring or pool in the mountains unknown to teamsters, so that when the team arrived it found nothing to eat. Still one can understand this. They stole the food that they might use it themselves. But what shall be said of their opening the faucets of the portable water tanks and allowing the water to run on the ground, so that the teams and drivers, well-nigh perishing with thirst after a hot day's journey, arrived to find them empty? Not only was this done, but pipes that led from springs up on the mountain side down to tanks on the trail and the tanks themselves were destroyed.

The company never prosecuted them for this. The manager said he did not dare to do so. He could have convicted one, now and then, by placing spies on the road, but he was afraid of what the rest would do should one be sent to prison. They were all in league together. Even on the short trail of 9½ miles between the borate mine in the Calico Mountains and Daggett, where there is but one way station, the water has been run to waste by these desert tramps.

The grub-stake eaters furnish a curious study in human nature. They are of all ages, sizes, and complexions, and of all temperaments and dispositions. Some are suave and insinuating, others brusque and insolent, even to their benefactors; but they are alike in indolence and a desire to live at the expense of some one else. That men should settle down to live on an island of the South Sea, where all the creature comforts are to be had for the taking, is not a matter of surprise. One may even find some sort of motive for the tramp who begs his way across the continent, stealing rides, betimes, on the brake beams of flying express trains. He has at least a life of excitement. But here is a man—a type—whose sole desire is to wander off across an arid waste; seeking nothing, seeing nothing, doing nothing; spending his days and his nights alone, with no shelter or defense against inclement weather or venomous reptiles, and no remuneration or reward beyond the bacon, the beans, and the flour which form his grub-stake. Sooner or later, through indecision or indolence, he is overcome by the heat or the lack of water, he wanders off the trail toward the springs, the mirage pictures before him, and so perishes where there is none to give him burial.

It is said that the desert quickly turns the brains of some men—makes them monomaniacs, so that once they have made a journey across it they become fascinated and return to it again and again, as an opium eater to his drug. On no other theory can I explain the existence of the desert tramp.

The stories of the wandering of the Gypsies are many, but I was surprised to learn that they are not unknown on the desert—surprised, because I could think of nothing to attract them there. Still there are attractions for the rovers. All the Arabs of the desert—both Piutes and whites—have little bunches of horses, while at the scattered mining camps are men easily cajoled out of hard-earned dollars by the dark-eyed fortune tellers. The victims are widely scattered, but because of their isolation are the more easily swindled. Every year, sometimes twice a year, a band of them crosses the Death Valley region, en route from Lower California to the Mormon towns of Utah.

CHAPTER VI.

SPORTSMEN IN THE DESERT.

DEATH VALLEY can hardly be recommended as a sportsman's resort. It is not a hunter's paradise. But because every true sportsman has a love of nature and is more or less of a naturalist, one in search of an outing might do worse than make a journey to this desert locality.

When in the course of my journey across the desert I had reached the old borax works in the Amargosa Valley, I found there a watchman named Leander Lee, who, for more than twenty years, had lived in the desert region of which Death Valley may be said to be the heart. In the conversations which followed my arrival, he told of his experience when hunting, prospecting, and even "holding little bunches of cattle" in the various oases of the region, in a way to excite the greatest interest.

Those who have read stories about the terrors of Death Valley with its poisonous exhalations—or even the stories that somewhat truthfully portray the conditions sometimes existing there—will doubt the possibility of finding any sort of game anywhere in the region, but the fact is that life is not always a burden there for either man or beast.

The Funeral Mountains that rise on the east side of the valley seem to be quite destitute of verdure when seen from the valley. Nevertheless, if one will climb to the higher altitudes he will find water in natural tanks, a few springs, and in places a considerable amount of bunch-grass. It is a volcanic region, but the lavas have partly decomposed. There are, moreover, many strata that were plainly deposited by water before some giant convulsion threw them up. Bunch-grass is spreading all over the country, the Arabs say, with wonderful rapidity.

Nourished by the grass, mountain sheep have been found there in considerable numbers. Indeed, during one month in the fall of 1891, the Indians killed thirty sheep on the peaks just to the north of Furnace Creek.

Their preparations for this slaughter very nearly created a panic among the prospectors that traverse the trails of the desert whenever the weather will permit. These sheep find their feed on the benches and in the gulches of the mountain side, and while eating, it is said, they never look upward. But when they are alarmed in any way, they fly up to the top, and if there be a ridge there, follow it to the highest peak. Having observed this peculiarity, the Piutes build blinds on the ridge-top runways. They started in during the fall of 1891 to build a number of such blinds on crests overlooking several Death Valley trails. The prospectors who saw these blinds jumped to the conclusion that the Indians were building forts to guard mines of fabulous wealth, and for a general attack on the white navigators of the desert. The blinds were in all cases low, semicircular walls of stone. However, the Indians wanted meat instead of scalps, and when all preparations were complete, posted the best marksmen in the blinds, while the rest chased the sheep up to the slaughter.

The sporting reader who has no sheep head among his trophies can probably get one in the Funeral Range as readily and with as little discomfort as he can anywhere, if he will go in October or November. There is probably but one place where he would be as likely to get a head at all as he would be here, and that is in the Sawtooth Range in Idaho. The northwestern part of Wyoming was once famous as the home of the sheep, but while elk and moose abound there now, I was told when in Jackson's Hole, in October, 1891, that sheep were rarely seen, and that it was only after prolonged toil and hardship that a sportsman could hope for a shot. With the aid of a white guide and

RESERVOIR IN DEATH VALLEY.

two or three Indians, a man in the Funeral Mountains could get plenty of running shots within moderate range, and no sportsman could ask for anything better than that.

Nor is the sheep the only game to be found in the desert region. All the migratory birds pass that way in their flight, and since the ranch was established in Death Valley they have made it a stopping-place. And no wonder they do, for it is a most inviting oasis. The waters of the little brook called Furnace Creek have been carefully conducted by a ditch and pipes out of the cañon down to a fairly level stretch of ground, and there turned into a double pond a half acre or more in extent. From this pond the water runs through ditches over about thirty acres of land, the most of which has been sown to alfalfa seed. What with the imported trees and native grasses and plants, that flourish under the inspiring influences of moisture and heat, the oasis must seem particularly attractive to the passing wild fowl.

During my stay there the ducks came dropping into the pond and settling on the meadows at such frequent intervals that a wing shot might have had no end of sport. As it was, a tarrying prospector killed enough so that we had an abundant supply of them each day. But they did not come in large flocks. I did not see more than four at one time. Geese and brant are sometimes found there, but not often, while the big white swan is not unknown.

At other oases in the desert the number of wild fowl is very great. At Fish Lake, north of Death Valley, where there is a considerable body of water in the low ground all the year round and vast beds of tules grow, the number of mallards, widgeons, teal, butter-balls, and what-not is incredibly large. I saw flocks of thousands when visiting the borax deposits of the vicinity. They fill the air by day, and the sound of their gabble can be heard the night through. The banks of the ponds were everywhere covered with feathers, showing where prowling coyotes, foxes, and wild-cats had feasted by night, while by day great hawks, and sometimes eagles, were seen hovering around, and not without success.

But the most interesting fact noticed about the wild fowl of the desert was the trouble they had when they settled in certain waters there. Because water is scarce in the region the smallest ponds, even the crystallizing vats at some of the borax works, attract them. In some of the vats they will remain swimming about all night; if the night is cold they often remain longer, whether they wish to or not, for the crystals forms so rapidly at such times that when morning comes the unfortunate birds are so weighed down with crystals they can not fly. Although I did not see this, there is no doubt of the fact, for I was told of it by Mr. F. M. Smith, who discovered the Teels Marsh deposit, and others confirmed the story.

A similar fate overtakes the wild fowl that settles in certain natural ponds in the desert, which are thoroughly charged with carbonate and sulphate of soda, as are those at Keeler, Inyo County, Cal. The ducks in these ponds become laden with salsoda, so they can neither fly nor dive, and so become easy prey for the Piute sportsmen. There is no one in the world who enjoys his sport so well as a Piute does when gathering in crystal-laden ducks, unless it be a British nobleman sitting in a "hot corner" while his retainers drive fat pheasants within pot-shot range and load his guns for him.

The loading down of ducks with borax crystals led a bright young fellow into a speculation, from which, it is said, he made a lot of money. He learned that the borax preserved the meat from decay, without in any way hurting its flavor, and thereupon he made a preserving powder for the use of sportsmen who might want to send game from the wilds to their friends. It was composed chiefly of powdered borax, but he gave it a taking name, advertised it in the sporting papers, and, although he sold it at a high price, he found a big demand for it. Straightway others imitated him. As a matter of fact, simple powdered borax, as it is sold in any grocery store, will serve every preservative purpose as well as any of these high-priced goods, so extensively advertised, and at a fraction of the cost.

A game bird, which I very unexpectedly found in Death

Valley, was the quail. A man, called "Bellerin" Teck, started in to make a ranch there once, and carried in a few California quails. Teck abandoned his scheme before very long, but the ranch made by the borax people became a permanent improvement, and in the luxuriant growth of brush about its borders the quail has found a congenial home. Away from the protection of the water-nourished brush, a quail could not live through a single summer in Death Valley, but within the oasis, the dry, heated air is so far modified that life, though far from pleasant, is tolerable.

Though not game, the blackbirds and robins in Death Valley, in the winter, are so numerous as to interest every sportsman who sees them.

Naturally, the predaceous beasts of the desert come to the oasis at night. Lee said he killed five wild-cats and one lynx while visiting the ranch. What James Dayton, the ranchman, has done, I did not learn, for the reason that he had gone into Daggett to buy when I was in Death Valley. Coyotes are found there, but are less numerous, for some reason, than they are over in the Amargosa Valley to the east.

Among other animals, of which I saw signs or was told stories, were badgers, bats, mice, and rats; gophers, jack-rabbits, cotton-tail rabbits, skunks, and foxes. Lee, who had a deal of dry wit about him, and who came over into Death Valley with me, frightened the outfit into making beds near the horses, after we got down into the Death Valley cañons, by saying that the foxes there had the disagreeable habit of biting off the ears and noses of incautious campers. The foxes are of a gray color, and much smaller, he said, than the eastern gray fox.

The Indians had burned some acres of mesquite and brush along Furnace Creek in their hunting for rabbits and rats. The stuff was still burning when we passed. It was exasperating to see such destruction in a region so scantily supplied.

While at the ranch it was visited by a native sportsman—a little black, dirty Piute. He was dressed in cast-off clothing of white men, and was armed with a bow and three arrows. The bow was of juniper, backed with raw sinew, and the arrows were of reed, tipped with juniper. They were effective against rabbits, rats, and lizards, and so satisfactory to the Piute sportsman.

THE CHAHWALLA.

There is one lizard there not to be despised by the white man. It is called chuckwalla by the whites and chahwalla by the Indians. Some of them are large enough to weigh three pounds dressed. The Indians place them as caught between two hot rocks to roast. The whites dress them and broil them on the coals of a sage-brush root fire, or fry them in bacon fat. The meat is very much like that of a frog's hind legs. There is also a desert terrapin. I saw the shell of one that was six inches in diameter. There is very little meat about them, but everything is fish that comes to the Piute net, including the kangaroo rats.

Very likely these rats would taste good to a hungry white man, but they are not inviting on sight. There was formerly, when the Death Valley borax works were running and teams were constantly passing, a great colony of these rats at Mesquite Well in Death Valley. When the teaming stopped they migrated. They are a comical kind of beast. The fawn-colored body is from four to six inches long, and the muscular tail from six to nine inches. Where numerous, they will gather about a camper, sitting up on their haunches with pious gravity, their big ears cocked forward and their eyes intently watching his every motion, ready to go away with great flying leaps when alarmed. They became so tame in the teaming days that the men sometimes cut the tail from one to see him go tumbling when he tried to jump without his counterpoise.

A small kangaroo mouse can be found in the Adirondack Mountains, where they are called deer mice, because of their color.

The kangaroo of the desert, it is said, never drinks water. I could not help doubting this, for the reason that at Mesquite Well, and again at Copper City Spring, near the base of Pilot Butte, I found them dead in the water.

Of serpents, and tarantulas, and scorpions, and centipedes, Death Valley has a great abundance. Of gnats and strong-jawed gad-flies it has intolerable clouds. There is one snake peculiar to the desert region, which is there called the "sidewinder." It is a little rattler, from fifteen to eighteen inches long, that flops about from side to side, instead of crawling as reputable snakes do. Moreover, it has a horn one-eighth inch long over each eye, and its bite is more frequently fatal than that of any other snake of the desert. Worse yet, it is so small that the warning ring of its rattlers is rarely heard, and the first knowledge the unfortunate victim has of its presence is when he feels the fatal sting of its teeth.

Whatever may be said of the habits of snakes elsewhere, they travel at night on the desert, and have a sleep-destroying habit of crawling into bedding.

Supt. Perry, of the Pacific Coast Borax Company's works at Daggett, while looking for a short cut over the mountains east of Death Valley, nearly perished on the desert through meeting a snake that did not bite him. On his way from what is known as Leach's Point, in a branch of Death Valley, over the mountains, one of his horses took sick. For seven hours he worked in the broiling sun to save that horse, but it died, and there he was afoot on the desert and more than fifty miles from the nearest water he knew of. He had a workman with him, but that was a disadvantage, for after the horse died they emptied the water from the keg into a bucket and found only enough to fill a small canteen and half the bucket.

At nightfall, though tired with the work of the day, they started on foot toward Hidden Springs, hoping to reach a small wagon that stood by the trail something like twenty-six miles away in Windy Gap, en route to the springs. It had been a scorching hot day. It was still extremely hot by night, but they toiled on, barely taking one small swallow of water at a drink, because the supply was so short. They had indeed not taken more than two such drinks when Perry heard the ominous whirr of a rattler beside the trail, and sprang to one side just in time.

By jumping, he escaped the rattler's fangs, but a fate almost as awful as the snake's bite overtook him. In his jumping he landed on some jagged rocks, and falling, spilled about all the water. They had just one pint left in the canteen, and Hidden Springs lay forty miles away.

The trail lay across the foot of Death Valley. If they succeeded in passing the valley and reaching the shade of the old wagon before sunrise, there was one chance in a hundred that they would live the day out and be able to walk to Hidden Springs the next night. On that chance for life they toiled on, and just after the sun had climbed above the mountains, they moistened their lips from the canteen and lay down under the wagon. But there was no sleep. They were tortured by the heat and thirst, and having no water to drink, and only enough to moisten the tongue so that it would not swell beyond their lips, they could only lie absolutely still and wait on the creeping hours.

"What does a man think of when dying thus of thirst?" said I to Mr. Perry. He replied:

"I remembered once seeing an old darkey drive into a town in Missouri with a big load of water-melons for sale. His horse was ebony and his rig ready to fall to pieces, but those great cool-looking melons were piled high in the old wagon bed. I saw the melons all day long, and imagined myself breaking them open and tearing out the juicy red heart and putting it dripping cool into my mouth."

Mr. Perry thinks he would have walked that night twenty-five miles to Hidden Springs. Perhaps he would, and perhaps he wouldn't. Had he not lost the water when the snake struck at him, he might have done so, but since it was lost there is a good deal of doubt about his strength being equal to the task. But he did not have to make the effort. Just before night a team came along, a day ahead of when it was expected. That night Mr. Perry set up until 1 o'clock to drink water.

On the whole, neither Death Valley nor the Mojave

Desert in general, is to be called good hunting-grounds, even though about 150 different birds and animals, all told, can be found there, but if the sportsman would like to kill game in a region where, so far as I could learn, no sportsman, save Government scientists, has ever penetrated for sport, let him try the Funeral Mountains. He will find the experience novel, and the scenery-part of the show worth the price of admission.

CHAPTER VII.

FREIGHTING ON THE DESERT.

HE "largest, most capacious, and most economical wagons ever built were manufactured on the Mojave Desert, for use in Death Valley."

The tourist among the deserts of Nevada and California will hear a good many curious statements from the scattered population he will find there—the one quoted above among the rest—and if he have any interest in horses or teaming, he will find the subject of desert transportation worth inquiry. There is probably nothing like it in all the world.

I got my first glimpse of desert transportation at the Nevada Salt & Borax Co.'s works, at Rhodes' Marsh, on the Carson & Colorado Railroad, Esmeralda County, Nev. The works for producing borax from the crude material, found in the marsh there, used nut-pine as fuel, and the wood was cut on a mountain-top, twelve miles away, piled up on a bench at the head of a cañon, and drawn thence in wagons to the works. One of these wagons was standing empty in a wood-yard when I visited the marsh, and, although not the largest in use, it was a sight to make an Eastern teamster gasp. The tops of the wheels came just level with the eyes of a tall man.

Over the divide at Teels' Marsh, some nine miles away, I found more wagons of the same kind, and, finally, down at the mining-camp of Candelaria and the little village of

Columbus, where there is another borax marsh, I saw what they called wood-trains—all loaded—trains, so to speak, of two great wagons coupled together and piled high with wood.

The woodsman of the East counts his load great when he has piled two cords on the easy-running bob-sleds in winter-time, but here the wood-hauler piles from five to six cords on each wagon, couples two of them together, and draws the train down the rocky defiles and winding cañons of the mountain-side and across the sandy plains, where the wheels of an ordinary Eastern farm-wagon, with its load, would cut in six inches deep.

Of course, no one pair of horses, nor any combination of horses, known to Eastern teamsters, could move, let alone haul, such a load. The swell young gentlemen who handle the ribbons over two pairs of horses, in front of a Newport coach, and the dignified driver guiding four pairs of heavy grays before a New York City safe truck, think themselves drivers of rare skill, and so they are. But the fuel-hauler of the desert commonly drives twelve horses, with the aid of a single rope in place of reins, and never has less than ten before him.

And yet he is but "a raw-hide" driver, when compared with those who had charge of the Death Valley borax teams.

When, in 1883, the manufacture of borax was first undertaken at the marsh in Death Valley, one of the best-known men in the desert region was Charles Bennett. He had taken up a claim on an oasis in the Pahrump Valley, in Southern Nevada, and had made a ranch of it that he afterward sold for $20,000. Here he lived, hundreds of miles from the nearest town, with the Piutes only for neighbors, unless, indeed, the scattered white Arabs of the desert—renegade whites and squaw wives—and one or two white

VIEW OF MOJAVE.

families, who lived at springs, from twenty to 100 miles away, could be called neighbors.

But in spite of this curious taste in the selection of a home, Bennett thrived on his ranch, and accumulated plenty of horses, mules, and cattle, with money in the bank at Los Angeles, through furnishing supplies to prospectors and trading with the Indians. He learned about the doings in Death Valley, and before the fire was built under the pans, had made a contract to haul the product over the desert to Mojave Station, on the Southern Pacific Railroad, as well as to freight the supplies from the railroad to the workmen in Death Valley.

Before the end of the year, when his contract expired, the company making the borax concluded they could do the freighting more satisfactorily with their own teams than by contract, and, accordingly, J. S. W. Perry, now superintendent of the Pacific Coast Borax Company's borate-mines in the Calico Mountains, and who had before that been employed in Mojave in the borax business, was put at work organizing a system of transportation over the desert, which should be adequate for the safe handling of all the product of the Death Valley region.

Some of the difficulties in the way of carrying out the company's plans may be mentioned, but scarce described so as to be fully comprehended by one who has not seen the desert to be crossed. Between Mojave and the valley proper there were but three springs of water. The road from the railway station led away over the sandy plain, in an easterly direction, toward a peak locally known as Granite Mountain, but called Pilot Butte in the reports of the California State Mineralogist, and by the early prospectors as well. It was just 50½ miles across this desert—a desert where the sand-laden wind forever blows, and the sun pours down with intolerable fierceness in summer—to the first spring, which was called Black Water. Beyond Black Water, 6½ miles away, was Granite Spring, at the foot of Pilot Butte, and the next spring was Lone Willow, twenty-six miles away, at the foot of one of the peaks of the Panamint Range. These last two spaces between springs were comparatively short distances between waters, but the next dry space was worst of all, for it was fifty-three miles to Mesquite Well, near the lower end of Death Valley.

And yet experience had demonstrated that a loaded team could only travel from fifteen to seventeen miles in a day. There was, of course, but one way in which those fifty-mile stretches could be crossed, and that was by hauling water for men and animals for the three days required in the passage between springs. Nor was that all. The desert does not produce a mouthful of food of any kind. Grain and hay had to be hauled as well as water.

There were other obstacles along the trail. It is a mountainous country. The road leaves Death Valley by what is known as Windy Gap. This gap is really what is known in that country as a wash. It is the bed of torrents that come pouring down after a cloud-burst on the mountain top. Volumes of water, in foaming waves twenty feet high, are said to be common enough, and others much higher are told about by the white Arabs. When a wave has passed, boulders are found scattered in all directions, gullies are cut out, and at the best only a bed of yielding sand is found for the wheels to roll over. Worse yet, this bed of sand rises on an average grade of one hundred feet to the mile for forty miles, while the grade for short distances is four times as much.

The entire length of this desert road between Death Valley and Mojave is 164½ miles. There was, of course, in all that distance no sign of human habitation. In case of sickness, accident or disaster, either to themselves or the teams, the men could not hope for help until some other team came along over the trail.

The first thing done by Mr. Perry was to obtain, by inspection or correspondence, the dimensions of all varie-

BORAX FREIGHTERS' CAMP.

ties of great wagons used by Pacific coast freighters. With these and the load carried by each wagon spread out before him, he proceeded to design the wagons.

The task he had set for himself was the building of ten wagons so large that any of them would carry at least ten tons. The reader who is familiar with railroads, in fact any reader who has traveled at all by rail, must have seen these legends painted on the sides of freight cars: "Capacity 28,000 lbs." "Capacity 40,000 lbs." (rarely) "Capacity 50,000 lbs." With this in mind, consider that these wagons for hauling borax out of Death Valley were to haul ten tons, or half a car load each—that a train of two wagons was to carry a load, not for one of the old-style, but for one of the modern, well-built freight cars, and carry the load, too, not over a smooth iron tramway, but up and down the rocky defiles and cañons of one of the most precipitous mountain ranges in the world, the Panamint. Because these were probably the largest wagons ever used, and because they were and still are completely successful, space may well be given to their dimensions in detail. They were as follows:

The hind wheel was seven feet in diameter, and its tire was eight inches wide and an inch thick. The forward wheel was five feet in diameter, with a tire like that on the rear wheel. The hubs were eighteen inches in diameter by twenty-two inches long. The spokes were made of split oak, 5½ inches wide at the butt, and four inches wide at the point. The felloes were made double, each piece being four by four inches large in cross-section, and the two being edge-bolted together. The forward axle-trees were made of solid steel bars, 3¼ inches square in cross-section, while the rear axles were 3½ inches square. The wagon beds were sixteen feet long, four feet wide, and six feet deep. The tread of the wagon—the width across the wheels—was six feet. Each wagon weighed 7,800 pounds, and the cost of the lot was about $9,000, or $900 each.

It is worth while to once more compare these wagons

with the best modern freight car. The best freight car for use on a steel track weighs 27,000 pounds, and carries a load of 50,000 pounds. Note that the car weighs more than half the load. Two of these Death Valley wagons very often carried 45,000 pounds, and sometimes 46,000 pounds of cargo, exclusive of water and feed for men and team, while their combined weight was but 15,600 pounds,

ONE OF THE BIG BORAX WAGONS.

or about one-third of their load. Moreover, all of the ten were in constant use for five years without a single break-down. The works in Death Valley were then closed down, but two of the wagons have been in constant use since, and are at this date (1892) running from the Borate Mine in the Calico Mountains to Daggett Station on the Atlantic & Pacific Railroad, where they bid fair to have an experience equal to that of the wonderful one-horse shay.

The building of the wagons was but the beginning of the work, though it should be said here that the building was all done in Mojave Village by men working by the day—it was not a contract job. While the wagons were building, the road had to be divided up into what might be called days' journeys. The heavy loads were to be brought in from Death Valley, and since only supplies for the workmen were to be carried out, the wagons would have but light loads one way. Of course the teams would not travel so far in a day with a full load as with a light one. Moreover they could not travel so far on the long up-grades, like that in Windy Gap, as they could down the long grade from Granite Spring toward Mojave. So the matter was figured over, and ten stations were established at intervals along the whole route, where the teams could stop for the night when coming in loaded to Mojave, while certain other stations were established for resting places on the way out to Death Valley, these last being located with a view of making a team travel further when light than when loaded.

So far as possible these stations were established at the few springs found along the route. Elsewhere dry camps had to be made. Here the natural lack of water was overcome by a system of wheeled water-tanks, very much like the tanks of street sprinklers. These were made to hold 500 gallons each, and were towed by the teams from the springs to the dry camps, and from the dry camps back to the springs to be filled again when empty. They were necessarily made of iron, because a wooden tank would dry out and fall to pieces when partly empty.

Then, in the language of the desert Arab, the springs were developed. Some holes were cleaned out and enlarged. At others that were not easily accessible from the best trail to be followed by the wagons, pipes were put in and the water run down to convenient tanks. At all tne stations from two to four feed boxes were built of lumber, each large enough to hold four bales of hay and six bags of barley, barley being the grain used on the desert as oats and corn are used in the East. The teams bound out to the valley filled the feed boxes, and then emptied them coming in. The greatest distance made by a team in cool winter weather, on a down grade with no load, was twenty-two miles. The shortest run for hot weather was about fourteen miles.

But it should be said here, that for the three months in the heat of the summer, from the middle of June until the middle of September, no teaming could be done at all. It was not possible for either man or beast to stand the terrific heat of even the Mojave Desert, not to mention Death Valley.

The teams consisted of eighteen mules and two horses. As was said, the man who handles four trained horses before a society coach, or eight huge Percherons before a safe-carrying truck, may think himself a pretty good driver, but in the desert, to use the desert term, he would be a sick raw-hide beside the man who steers eighteen mules with a jerk-line. To compare the one with the other is like comparing a Corinthian yachtsman, or the deck-hand of a harbor scow, to the captain of a Black Ball liner, if we may use a nautical simile in a story of the desert.

In building the desert freight train, the front wagon receives a tongue of ordinary length, while from the rear axle projects a little wrought-iron tongue about three feet long. The second wagon has a tongue, say six feet long, with a stout vertical ring on the end of it, which, when the two wagons are coupled together, slides over the three-foot tail of the front wagon. Then, to hold the two wagons together, a stout chain runs from the front axle of one to the front axle of the other.

The horses and mules are harnessed up in pairs. The horses are attached to the wagon at the tongue, and a great, handsome 2,800-pound team it is—gentle, obedient, and strong as a locomotive. Ahead of them stretch the mules, their double-trees geared to a chain that leads from a forward axle. The most civilized pair are placed in the lead and the next in intelligence just ahead of the tongue, while the sinful, the fun-loving, and the raw-hides fill in between. The nigh leader has a bridle with the strap from the left jaw shorter than the other, and from this bridle runs a braided cotton rope a half an inch in diameter, through fair-leaders on each mule to the hand of the driver, who sits on a perch on the front end of the wagon box just eight feet above the ground. That rope is known as the jerk-line, and its length is not far from 120 feet. The team that draws the desert freight train stretches out for more than 100 feet in front of the wagon.

If historians and poets have been justified in writing rapturously about the Arab and his steed, what may we not say of the Death Valley teamster and his mules? To see him soar up over the front wheel to his perch, tilt his hat back on a rear corner of his head, gather in the slack of a jerk-line, loosen the ponderous brake, and awaken the dormant energies of the team with "Git up, —— —— you; git up," is the experience of a tourist's life-time. And when at the end of a journey, the teamster pulls up beside the dump with the mules in a line so straight that a stretched string would touch the ear of every mule on either side of the chain, as has often been done, one wants to be introduced and shake hands, as with "one whom lesser minds make boast of having seen." And when one sees the mules settle forward in their collars, feeling gently of their

LOADED TRAIN LEAVING DEATH VALLEY.

load, until at last the chain stretches as firm as an iron bar, and with one accord start the train of well-nigh 60,000 pounds weight almost as though it was naught, he wants to be introduced and shake hands with the mules, too—that is, figuratively speaking. Their intelligence is such that he would be proud of a speaking acquaintance with them, but if he knew the mules he would be a little shy about getting within hand-shaking range.

It is wonderfully interesting, too, to watch the mules as they turn a sharp corner in a cañon, or on a trail where it rounds a sharp turn on the mountain side. Span after span, near the end of the tongue, often without a word from the driver, will jump over the long chain and pull away on a tangent that the heavy load may be dragged around. Even then the novice wonders how they succeed, for some of the curves are so sharp that the leaders pull in one direction while the wagons are traveling very nearly in an opposite one.

In their short journey after fuel, the drivers of the ten-horse teams often manage their outfits alone. It is but a day's trip from the village to the wood camp and back; but in freighting over the desert with a twenty-animal team, every driver has an assistant called a swamper. The swamper's duties are multifarious. On a down-grade, he climbs to a perch on the rear wagon and puts on the brake; on the up-grade, he reasons with and throws rocks at the indolent and obstreperous mules. As meal-time approaches he kicks dead branches from the grease-brush along the route, and pulls up sage-brush roots for fuel. When the outfit stops, he cooks the food while the driver feeds the animals, and when the meal is over, washes the dishes, which, with the food, are carried in a convenient box in the wagon.

The mules get their grain from boxes which are arranged to be secured to the wagon tongue and between the wheels,

READY FOR A SWING.

when feeding. They eat their hay from the ground. Beyond feeding and watering, the animals get no care—they curry themselves by rolling on the sand, and rolling with cyclonic vigor, at that. The cloud of dust raised when an outfit of mules starts in for a lark is suggestive of a Death Valley sand-storm, and there is nothing to compare with their cries of glee after the rolling is done. The work is not wearing on the animals. It is common and polite to say to a driver, when a thin or scrawny mule is seen in a big team: "Been getting a raw-hide, hey?" which, being interpreted, means: "Ah, I observe you have recently purchased an animal unaccustomed to the work."

Quite as interesting as the teams and the freight trains of the desert are the men who handle them. The drivers receive from $100 to $120 per month, and the swampers about $75. They furnish their own food and bedding. The bill of fare served at a desert freight camp includes bacon, bread, and beans for a foundation, with every variety of canned goods known to the grocery trade for the upper strata. They carry Dutch ovens for their baking, pans for frying, and tin kettles for stewing. On the whole, however, they do not eat much fancy canned stuff, and a cobbler made of canned peaches serves for both pie and cake.

"We don't care much for gimcracks, but we're hell on grub. The gimcracks don't stay by ye," as one said. They rarely carry liquor for use on the road. I observed that empty bottles on some of the desert trails were as thick as good resolutions on the road to sheol, but the teamster did not empty or leave them there. They had served to cheer the road for gentlemen en route to inspect Breyfogle, Gunsight lead, and Peg-Leg mines, discovered by enthusiastic eaters of grub-stakes.

This is not to say, however, that the teamster is a disciple of Neal Dow, or the Woman's Christian Temperance

Union. While the five trains were running regularly between Death Valley and Mojave, the chief care of Superintendent Perry was to keep them moving regularly. He had the road so divided that the teams went out to the valley, got loaded, and returned to Mojave on the twentieth day at 3 o'clock with a precision that was remarkable. At Mojave the teamster was allowed to have the rest of the day and night to himself, and it usually happened that when the hour of starting came next day, he rolled in instead of soared to his perch, and then, as he blinked his eyes and pawed the jerk-line, said:

"Git hep-th-th-th-th yougithop."

It is a matter of record that the mules understood him, nevertheless—that, in fact, these long-eared, brush-tailed tugs of the desert never did but once fail to understand the driver, no matter what his condition. On that occasion the driver, instead of getting drunk, had gone to hear an evangelist preach, and had been converted. Next morning, it is said, when he mounted the wagon and invited the team to go on, the mules, with one accord, turned their heads over their shoulders, cocked forward their ears and stared at him. He had omitted the customary *emphasis* from his command.

It is a curious fact—a fact that a thoroughbred Kansas boomer will scarcely believe in—that the building of a railroad to a desert mining camp invariably decreases the life and activity seen on the streets and among the business houses. The railroad benefits the mine owners, but injures everyone else. The explanation is simple, however. Before the railroad reaches the active camp, all the supplies are brought by teams, and so are the mails and the passengers. When the railroad comes, the teamsters and swampers drive away to return no more, and the railroad brings none to take their place. In fact, it would take a pretty lively citizen to fill the place of a departed teamster, in any event.

"There was a faro bank running most of the time at Mojave. It was a good thing for us, for the teamsters could go broke in one night and be ready to go out over the road in the morning," said Supt. Perry.

That was by no means a heartless remark, as it seems to be at first blush, for if the teamster did not gamble away his money, he was sure to get drunk and spend it in ways more harmful, while if by any chance he got the wages of two months in his pocket at once, he would rush off to Los Angeles for a spree that would take a fortnight or more to recover from. The teamsters are, with rare exceptions, unmarried men.

The life of a teamster on the desert is not only one of hardship, it is in places extremely dangerous. Mention has been made of the grades up which the loads must be dragged. There are other grades down the mountains, like the one, for instance, on the road from Granite Spring toward Mojave, where the plunge is not only steep, but the road-bed is as hard as a turnpike. The load must go down, and so when the brink is reached the driver throws his weight on the brake of the front wagon, the swamper handles the brake on the rear one, and away they go, creaking, and groaning, and sliding, till the bottom is reached. If the brake holds, all is well, but now and then a brake-block gives way, and such a race with death as then begins can not be seen elsewhere. With yells and curses, the long team is started in a gallop, an effort is made to swing them around up the mountain-side, a curve is reached, an animal falls, or a wheel strikes a rock or a rut, and, with thunderous crash, over go the great wagons, and the teamster who has stuck to his post goes with them. There are many graves on the desert of men who died with their boots on,

ON THE DOWN-GRADE.

but some of them hold men who were killed while striving to guide a runaway freight-team in a wild dash down the side of a desert mountain.

As one may suppose, the effect of desert life upon the teamsters is almost every way deteriorating. The men who drove from Mojave were out twenty days for each half day in the settlement, and the settlement itself was but a collection of shanties on as arid a part of the desert as can be found outside of Death Valley. They were not men of education or very wide experience. Their topics of conversation were few. The driver and his swamper had very little to say to each other. To all intents and purposes each lived a solitary life. Being thus alone they grew morose and sullen. Their discomforts by night and their misery by day in the desert heat added to their ill nature. They became in a way insane. It was necessary whenever a team came in to inquire of each man separately whether he was perfectly satisfied with the other, and whether a change was desired or would be objected to. If the least ill will was displayed by one toward the other, a new swamper was provided, lest a fight follow on the desert and one kill the other. Even the greatest precaution could not prevent murder. The soil at Saratoga Springs, in the Amargosa Valley, is stained with blood, a human corpse once swung from a telegraph pole in Daggett, and a rounded pile of stones in Windy Gap is marked "Grave of W. M. Shadley," all because human flesh and human brain could not endure the awful strife of life on the desert. Because these are phases, and illustrative phases, of life on the desert, the stories of these crimes should be told.

Fortunately the stories are but brief. A team was coming in to the railroad from the borax works in the Amargosa Valley. At Saratoga Springs they stopped for the night. There the teamster and swamper quarreled, and the swamper hit the teamster on the back of the head with a shovel, as he sat by the camp fire, killing him at once.

Then the swamper buried the body close by the spring and lay down to sleep by the grave. In the morning he hitched up the team and started to drive in. But he was no teamster, and soon had the mules in a tangle, and the wagons, big as they were, overturned, the fall breaking the swamper's leg. In this condition he crawled about among the animals and turned them all loose save one horse, which he somehow mounted and rode away over the long, hot divide, with the broken limb swinging about and the broken bones grinding together, till he reached the works once more.

His terrible condition and untrue story of the trouble with the teamster awakened the deepest sympathy—a feeling which lasted until he had been sent in a buckboard, a journey of 105 miles over the desert, to a surgeon. When the workmen came to dig up the body of the teamster, that it might be removed to a healthful distance from the spring, they found he had been foully struck from behind, and they wanted to lynch the murderer. But they did not do it, and because of the discomforts and dangers of a trip over the desert, neither the coroner or the district attorney of the county would investigate the matter.

Daggett's only lynching was due to the murder of a teamster. His swamper, for some fancied wrong, was moping about the village, drowning his care in liquor. Another teamster advised him to kill the offender. Early next morning someone passing the blacksmith shop heard groans behind it, and there was found the offending teamster alive, but with his skull crushed. Beside him lay one of the huge spokes used in building wheels for desert wagons. One end was covered with blood and the hair of the dying teamster.

VIEW OF DAGGETT.

Two nights later, when it appeared that the Justice was about to turn the swamper loose for want of direct evidence of guilt, a masked mob took both the swamper and the teamster who had advised the crime, from the lock-up. The telegraph poles at Daggett have a single cross-arm. Two ropes were thrown over one of these arms, and nooses in the ends were put about the necks of the two prisoners. Both men had until this time thought the movement a bluff to frighten them into confession. Now they would have begged for mercy, but before the trembling lips could gasp half a sentence the tightening ropes lifted them from the ground.

However, it was really but a bluff on the teamster. He was soon lowered to the ground and advised to leave town. He left. The swamper now "holds down a six-foot claim on the mesa," just beyond the village limits.

CHAPTER VIII.

A NOVEL ROAD.

THE natural wonders of Death Valley have probably been more minutely and extensively described by professional writers than any other spot they never saw, but an artificial wonder there has some way escaped these untraveled scribes. The old borax works there were built on the east side of the valley, a couple of miles or so above the mouth of Furnace Creek cañon. The road thence to the railroad led down the east side of the valley for several miles, and then had to cross over to the west side, because no water to speak of can be had on the east side below Furnace Creek. Moreover, the land on the west side lay much better for a road. But how to get the wagons across the valley was a problem. From end to end, the center of the valley is one long salt marsh, and in most places it is so soft or wet that even a man would need snow shoes to insure his safety. Elsewhere, however, the ooze has been crusted over. This crust is, in places, very thin and treacherous, and only in one locality does it seem to be firm. Wherever this crust has been cut through, the slimy salt mud has been found to be of immeasurable depth—unmeasurable with any line or pole. Dr. C. Hart Merriam's corps of scientists cut through in one place and easily shoved a pole down fifteen feet. There is no guessing how much deeper the slime was.

However a road must be had, and so the workmen went

about over the marsh where the crust seemed to be thickest and sounded it with sledge hammers. They found the crust was a mixture of salt and sand. Eventually, a route was decided upon. The road was then to be graded, and probably for the first time in the world a road of the length of this one was graded exclusively with sledge hammers.

It was made across a stretch of solid salt some eight miles wide. In a sense, it was level—there were no hills or valleys. In another sense there was scarce a level square inch on the whole bed, for the salt crust had, probably through the influences of heat and moisture from below, been torn and twisted and thrown up into the most jagged peaks, pyramids, and cris-crossed ridges imaginable. They were not high—none, perhaps, more than four feet—but there was not even level space for a man's foot between them. Every step made by the explorers was on a ragged point or edge of some kind. The nearest approach to anything like that salt bed I have ever seen was on the ice on Lake Erie, where two fields had been jammed together by the wind and held so by the frost. The ragged ice masses were somewhat like these salt masses. They were larger, but they were neither so sharp nor in any way so difficult to cross.

Judging that the crust would sustain the weight of the wagons, the workmen swung their sledge hammers day after day, until they had beaten down these pinnacles into a smooth pathway six feet wide. It was, perhaps, the most laborious engineering work ever done in the country, for the climate, and the location far from civilized habitations, combined to retard the efforts of the workmen.

As one enters the easterly end of this road, two unmarked graves are seen in the salt crust near the track. They are graves of unknown men who died there from the heat, and, after the fashion of the country, were buried where they fell—they were covered over with pieces of salt broken from the pinnacles near by; the crust was too hard to warrant digging into it. One must travel a long time to find two more graves like those, if, indeed, two more can be found anywhere in the world.

CHAPTER IX.

GATHERING DESERT FUEL.

OF ALL the features of life on the deserts of Nevada and California, none is more likely to attract the attention of a tourist, and particularly a tourist whose home is in the Eastern States, than the gathering of fuel for industrial uses. In the first place, it strikes the unaccustomed spectator as a novelty that any industry needing a fuel should be found in a desert of the kind, not to mention the fact that the desert should furnish the fuel supply needed. However, industries there are and fuel to keep them going. That the industries are all connected with mining of some sort scarce need be said—the mining and smelting of precious metals and the gathering and refining of borax.

The traveler on the desert commonly has his attention first drawn to the fuel subject the first night after leaving the railroad station, when he has to make camp and rustle for a fire to fry his bacon and boil his coffee. For the camper's purpose fuel is everywhere abundant throughout the desert regions of both Nevada and California, save in the alkali flats, which are absolutely barren. The mesas and valleys are everywhere covered with sage and grease brushes, on many mountains the nut-pine and in some places the cedar can be found, while Death Valley and some other localities have many groves, or rather thickets, of the mesquite tree.

To the stranger the pine, the cedar, and even the mesquite

trees look something like fuel. He would not call them on sight first-class, by any means, even for a camp-fire, because the trees are at best but overgrown bushes, rarely more than fifteen to eighteen feet high, with branches that sprout in all directions from the clefts in the rocks where the trees

GREASE-WOOD.

find standing-room; a log fit for the foundations of a fire such as one is accustomed to build in fishing trips through the woods of Maine, the Adirondacks, or the Canadian wilds, is as hard to find as a spring of absolutely sweet water. Nevertheless, big roaring camp-fires can be made if the tourist happens to be in the tree region. But on the mesas of the Mojave Desert the case is different. Grease and sage brush form almost his only resources.

The grease-brush, which is found on the Mojave Desert in greater quantity than any other, is a bush with many slender branches starting out in all directions from a common center at the ground. It is much like an uncared-for currant-bush in an Eastern garden. The color of the bark is a dark gray or brown, but the leaves are of a color to give the whole bush a yellowish-green cast. The branches, even at the base, are rarely larger than a man's finger, though they occasionally reach an inch or more in thickness.

Here, then, in these slender twigs must the camper on the desert find his fuel for his culinary arts. If rather discouraging at first sight to a camper, what must it not be to a man who wishes, say, to fire a set of boilers of more than 100 horse-power? Even when grubbed up by the roots, which are short and thick, the grease-bush fuel burns up with a flash and is done. For the camper a good quantity of live coals remains, but for the manufacturer who would run a big boiler the coals are of little use.

My first satisfactory view of the gathering of brush for fuel for an industrial use was after crossing the borax marsh, bound south, at Columbus, Nev. I was driving along with Supt. Chris. Zabriskie, of the Pacific Coast Borax Company's works at Columbus, when a wagon drawn by two horses, and loaded ten feet high with something, appeared some two or three miles away in the gap leading through the mountains toward Fish Lake.

"There's a load of desert hay," said Zabriskie. "It makes better feed than you would suppose."

I noted carelessly that the load was very high and well built up, and turned to look, as I had been doing, at the wonderful colors of the bare mountains to the east. After a little I looked again at the "hay." It was close at hand then, and behold it was a stack of sage-brush equal in size to a two-ton load of timothy fresh from the field. A weazen-faced driver sat down close under the front of the load. At a word he stopped, with a look that showed somewhat of apprehension, while I snapped a camera.

LOAD OF DESERT HAY.

The brush was a load of fuel en route to the borax works on the Columbus Marsh, where it was to be used under the open pans in which the crude material is there treated in the process of refining described elsewhere. It had been gathered on the desert, some eight or nine miles from the works, the brush within that radius having all been burned off. Both at the Columbus Marsh Borax Works and at the works at Teels' Marsh I saw this brush used as fuel. It was dumped in great heaps handy by the mouth of the furnace, and there pitched under the boilers by the pitch-fork full. Light and flashing as the fuel is, I noticed that the fireman was not obliged to keep constantly at work; about half his time was occupied in tending the furnace and half in leaning on the fork or sitting on a rock and gazing stolidly at the scenes round about.

Down at the works of the San Bernardino Borax Mining Company, on Searles Marsh, San Bernardino County, the desert brush was used for fuel for many years under a boiler of at least sixty horse-power. The desert from six to eight miles up the valley was stripped bare by workmen who gathered the brush in loads, and eight horses were used to draw them. Three two-horse loads of this brush were said to be equal to two cords of nut-pine wood, another desert fuel.

In one of the factories, if one may use the term, at Teels' Marsh the fuel used is nut-pine. The nut-pine grows at an elevation of 9,200 feet above the sea, on a mountain northwest of the works, the distance from the works to the wood-camp where the wood is piled for measurement being seven miles. Thence to the huts of the wood-cutters scattered among the trees is about two miles. As said, the nut-pine is a scraggy, overgrown bush. It sometimes grows twenty feet high in the deep, shaded cañons where the snow lasts until late in summer. Elsewhere it may average ten or twelve feet. It is scattered about over the mountains as sage or grease brush is over the plain.

As soon as the weather will permit in the spring, and long before the snow is gone, the wood-cutters, usually in pairs, go up among the trees and prepare for their summer's work. A little dugout is made in a cleft or beside a jutting rock. A low chimney of broken stone is built in one corner. A length of jute bagging serves for a door. A dry-goods case serves at once for a table and cupboard, while soap-boxes answer for chairs. The bedding, a few gray blankets, lies on the floor in one corner, and the floor is the ground picked free of stones.

It is a simple home - about as good as that of a Piute or a badger, but it is in a location that in one way might awaken the envy of the most cultivated and refined. Such pictures as lie spread out for the gaze of the desert wood-

WOOD-CUTTER'S HUT.

cutter can be found nowhere else in the world. People travel many miles to look off over the ocean, that they may feel if not comprehend its immensity, but here the tourist, at the hut of the wood-chopper, looks over a stretch of swelling buttes and rolling ridges so vast that the mind seems to fall but little short of comprehending the infinite. The wonders of the pictured rocks of Lake Superior, and of the tinted foliage of tree-covered mountains of the Hudson, have been told from time beyond memory, but the wood-cutter from his hut sees such vast piles of colored strata and rocks, such huge mountains of variegated and unchanging colors rising thousands of feet into the dreamy blue atmosphere, that words fail to describe their beauty. A wealthy citizen of the metropolis has spent half a million of dollars in buying land for a house-site in the mountains of North Carolina. I have stood upon the site, and it was to my mind well worth the money; but the penniless wood-cutter of the desert gets a site for nothing that is simply incomparably more magnificent. Let anyone who doubts this, visit a Nevada wood-camp and see.

Now and then, but rarely, one finds a wood-cutter who appreciates his location. It is a curious fact that men with parts of their minds fitted by nature and early cultivation for better things may be found among the wood-cutters, clad in overalls and jumpers, swinging the ax with grimy paws, and working their jaws the while over vile tobacco and viler profanity. As a rule the wood cutters are coarse-grained men who see only in the site the animal comforts of shelter and convenience to their labor. To them the best camp is the one where the snow-water remains longest with them.

When the melting snow is gone, water must be brought from below. It is at Teels' Marsh, as elsewhere, brought to the wood-camp in barrels placed on the wagons that draw the wood from the camp to the works, and from the camp it is carried up to the men at the dugouts in kegs strapped on the backs of burros. The melted snow is cold and sweet. The water that is brought up in barrels and kegs, on which the fierce sun beats continually, is anything but tempting as a beverage. It is not only warm, it is hot, and the wood-cutter has no use for warm water save in the coffee-pot. It is true that warm water would remove dirt better than cold water would do, but the wood-cutter does not want to remove dirt. There is a tradition that a man was once employed on the mountain above Teels' Marsh cutting wood who washed his hands and face every day; but he did not remain there long. The rest of the gang ostracised the dude.

Wood-cutting on the mountains is not hard work, as wood-cutting goes. The nut-pine is a peculiar wood. The cutting is chiefly done with the pole of the ax—the men break the limbs instead of chopping them to pieces. The wood is very brash. Indeed, the end of a broken dead stick looks very much like the end of a rotten stick. The fiber breaks across without even the slightest trace of a sliver. To a man accustomed to the maple, birch, hickory and other hard fuel woods of the East, the nut-pine looks like a rank imposition on the wood-buyer. But if one will open a furnace where it is burning he will find such a mass of flame as nothing short of sugar-maple could produce. It is a fuel that "stays by the furnace," as one fireman put it.

An ordinary wood-cutter can cut 2½ cords of nut-pine a day, and the price paid for cutting is $1.50 per cord. That sounds like big pay for a wood-chopper in the woods of New York or Pennsylvania, but it is not great for Nevada or California. The men board themselves at a moderate price, say 50 to 75 cents a day. They work till they get anywhere from $100 to $300 ahead, and then they spend it like lords in some town where creature comforts are cheap and bad. After that they go chopping again. Most of the cutters are Americans, with some English and Irish among them, and now and then a Mexican.

From the spot where the tree grew the wood is carried on mules or burros to the camp where it is piled and measured, and which is always located on a bench on the mountain-side, or in the cañon.

One of the marvelous sights of the Nevada mountains is the wood-laden mule en route to the camp. The Mexicans have almost a monopoly of wood-packing, as it is called. A wooden sawbuck saddle is cinched on the animal's back, over which are hung two round bars of iron, like big J-shaped hooks, that hang down on each side of the mule. The wood is piled into these big hooks and up over the mule's back until there is about twice as much wood as

WOOD IN THE RANKS.

mule in the outfit. In fact, each mule carries one-fifth of a cord of nut-pine, and is happy under the load. Having loaded his mules, the Mexican starts them, secured one to the other by long halters, down the precipitous side. The picture thus made is one never to be forgotten by the tourist. Writers of books on naval architecture, and other scientists, have much to say about centers of gravity and meta centers. They conclude that when the meta center is below the center of gravity in a ship she is sure to turn over on the slightest provocation. They teach that when the center of gravity of any body gets beyond the line of its base it is sure to upset. But they don't know the mule in the wood business on the steeps of the desert mountains.

The Mexican commonly receives $4 per cord for packing wood, say two or three miles, to camp. Here it is piled in great ranks. In one camp I visited I saw 300 cords piled ready to be hauled to the works in wagons. And the hauling in wagons is quite as interesting as any other part of the wood-gathering work; but the describing of the wood-hauling has been done in the chapter devoted to the freighting business of the American deserts. It is sufficient to say here that the wood is hauled down the cañons in loads of from ten to twelve cords each, with ten or twelve horse teams before the loads, and that the wood delivered at the works costs anywhere from $9 to $12 per cord. That at Teels' Marsh, Nevada, cost $9, and it was the cheapest wood I found in a journey of over 1,000 miles among the deserts. In the production of borax at Teels' Marsh no less than 1,500 cords of nut-pine are required every year, besides the sage-brush used in the outworks.

As was said before, nut-pine is the product of the highest altitudes. Mesquite, on the other hand, is found in places on the great desert, even below the sea-level. In my journey through Death Valley, I camped in a mesquite grove, at Mesquite Well, at least seventy-five feet below the sea-level. Mesquite grows in the cañons as well, but much more can be found in the low valleys than in the cañons. It is worth nothing that the mesquite will not grow without water—that wherever a mesquite thicket or grove is found, water, and as a rule good water, can be obtained by digging. This was a noticeable feature of Death Valley. All along the westerly side of the valley are great clumps of the mesquite, and back of each clump is a cañon that heads up in the snow-capped Panamint Mountains. The melting snow runs under-ground, but it runs nevertheless, for the soil is very loose and porous there.

The gathering of mesquite for fuel in Death Valley is commonly what Californians would call a placer-mining, or possibly a quarrying, proposition. They cut mesquite wood in Death Valley with a shovel and a mule, so to speak.

As was said, the mesquite tree grows on low ground. It is simply a great bush, low and scrubby, but rarely so gnarled as its high-placed compatriot, the nut-pine. Very many of even the large trees, say those six inches through at the butt, have a smooth green bark. It is a thorny tree, and when alive and healthy not pleasant to attack for any purpose. Moreover, the green wood does not make such good fuel as do the dead dry branches and trunks, and so the dead wood is gathered when fuel is wanted. Nor is that all that is gathered, for the roots of the mesquite are as large and wide-spread as the top, and more valuable as fuel, for they may be, and are, used in place of charcoal in desert forges.

As was said, this fuel is gathered with a shovel and a mule. The sand-storms of Death Valley, which are described elsewhere, bury out of sight grove after grove of mesquite trees. As the sand piles up about them a vain struggle for life begins. Shoots are thrown out on the side least covered by the sand. The top lengthens out, the free branches grow longer and more slender. One need have but little sentiment to enable him to see in this struggle something akin to that of a human being caught in the toils. Having no helping hand, the tree is at last buried alive by the relentless sand, and while yet the smaller branches are uncovered, it turns from bright green to a sickly yellow color, and then to the gray of death. Very soon the small branches drop off, and only a mound of sand remains, with here and there a black stub projecting, to tell the story of what is buried within.

Now comes the fuel-gatherer with his shovel and his mule. A few jabs in the sand uncovers a tree trunk or root, and then the man takes two hitches with a chain about the old stub, howls with familiar profanity at the mule, and in an instant out comes a stick of wood pleasing to behold. Mounds burying from four to six cords of wood are found there, and an active mule with a man "will take out a whole lot in a day," as my guide said to me when I asked about quantities. The Piutes, the Arabs of this portion of the American Desert, look with disfavor on the cutters of nut-pine, but are calmly indifferent to acts of the mesquite-wood miners. Both trees produce seeds much prized as Piute food. The pine-nut when roasted is not to be despised by cultivated tastes, even if it be a little like turpentine in flavor, while mesquite beans when roasted are fair to eat. But the Piute notes that the wood-choppers cut green as well as dead nut-pine trees, while only the dead mesquite is wanted for fuel. The one destroys a good crop-producer, a crop that enables the Piute to live all fall in comfort by the sweat of his squaws. Should the nut-pines ever be exterminated, the unfortunate buck will have to rustle somewhat in the fall as well as at other seasons.

A story about desert fuels would be incomplete were no reference made to the use of a fuel there that is imported from another region. The American desert is, perhaps, the last place in which one would expect to find it, nevertheless, hundreds of barrels of crude petroleum are consumed every year in one establishment in the Mojave.

The gathering of any kind of desert brush for fuel eventually becomes a pretty heavy drain on the profits of an industry, for the reason that the brush, being grubbed up by the roots, does not readily replace itself in the territory from which it was taken. At the San Bernardino Borax Works the gathering of brush eventually required the active services of nearly a score of mules and as many men. In this emergency Supt. Searles decided to substitute crude petroleum. The system adopted was very simple.

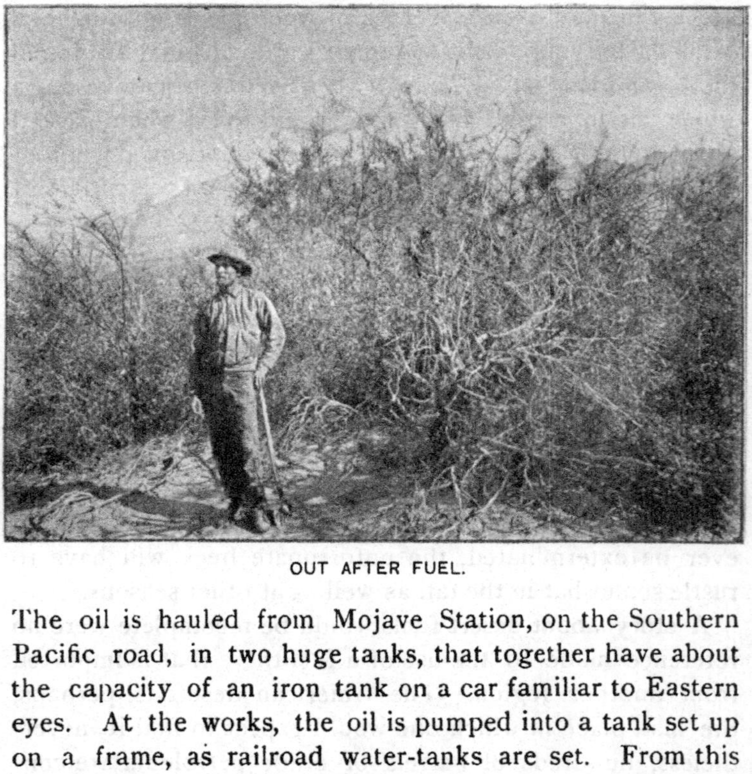

OUT AFTER FUEL.

The oil is hauled from Mojave Station, on the Southern Pacific road, in two huge tanks, that together have about the capacity of an iron tank on a car familiar to Eastern eyes. At the works, the oil is pumped into a tank set up on a frame, as railroad water-tanks are set. From this tank the oil runs through a slender pipe to each fire-box, there being two boilers in the works. In the fire-box the oil is combined with a jet of steam and fired, and a roaring hot fire is the result. It has been found that 120 horse-power for twenty-four hours is produced by a consumption of 683 gallons of crude oil. The cost of the oil amounts to about seven hundred and fifty dollars a month, besides the cost of transportation over the desert.

Although not used for fuel in any industry, the Yucca palm, that grows abundantly in some sections of the Mojave Desert, must be mentioned, for the reason that a curious transformation is said to take place in the wood, sometimes, after it has fallen to the ground. When cut green or when it has died naturally this tree has ordinarily a thin woody shell beneath the bark, and within the shell a soft pith. As a fuel, the tree would attract attention nowhere in the world save on such a desert as this. But occasionally, so the desert Arabs say, the trunk petrifies, and in the earlier stages of this change makes a fuel second only to the best soft coal. The soft wood becomes solid and brittle; and when fired burns with great and lasting heat.

CHAPTER X.

A NEVADA NARROW GAUGE.

EXPLORERS who wish to visit Death Valley and experience the hardships of a desert journey have a choice of two routes, one from the South and one from the North. In winter both are not alone practical; for a hardy traveler accustomed to camp life in the open air they are delightful. In summer, no matter which he takes or what his experience in outdoor life may have been, the tourist will wish he had taken the other. If he lives to return to civilized homes he will wonder how in the world he ever happened to take either. But there is one reason, if no more, why he should take the northern route at any season, and that is that he may see the Carson & Colorado Railroad.

In the journey made through the borate deposits of the Pacific Coast I first entered the great desert region in which all but one of those beds are found, in a car of the Carson & Colorado Railroad. The tourist who travels that way is not unlikely to wonder which is the more interesting, the road or the desert region. He is not unlikely to wonder, too, what the road was built for.

The Carson & Colorado is in one respect unique and in several respects remarkable. It does not start at Carson neither does it terminate in Colorado nor at the Colorado River. Its initial point is an arid mountain side, so arid, indeed, that water for use in the little settlement there has to be imported in big tank cars built for the purpose; the road runs thence 300 miles across a desert to terminate at a sal-soda lake in California. The settlement at one end is known as Mound House, Ormsby County, Nev. That at the other is Keeler, on Owen's Lake, Inyo County, Cal.

A passenger train, a freight train and a milk train run each way over the northern division (from Mound House to Candelaria) of this road every day, but, in order to economize, one engine is made to pull all three trains. The passenger train includes one coach and a combined smoker, mail, milk, baggage and express car. The freight is not unlike other freight trains save in length; it is a short train. I left Mound House in the coach one bright morning in November. There were five passengers in all, besides a newsboy, whom the conductor called Peanuts. Peanuts carried a bundle of San Francisco newspapers one day old, a basket of fine red apples, and a box of Chinatown cigars. By the time he had reached Cleaver, where we met the train bound north, he had sold two papers and one apple on the train and had smoked one of his own cigars. At Cleaver he boarded the other train and found not one passenger on it.

The train stopped for dinner at Wabuska, and Wabuska is a station located on an oasis. It is in a valley that is novel in the topography of Nevada, for it is well irrigated. Here we annexed the milk train—perhaps it would be more accurate to say annexed the milk outfit. The ranchers in Wabuska run somewhat to the dairy business, and supply the villages and camps along the route as far as Candelaria, about 100 miles away, with fresh milk. Milk cans are a familiar sight in the East and particularly to the people of the metropolis. The tourist from the East has therefore no difficulty in recognizing the milk cans on the Carson & Colorado Railroad as milk cans, particularly if his eyesight is good, for they are shaped like Eastern milk cans. The milk cans of the East hold from ten to thirty gallons. Those on the Carson & Colorado Railroad hold from one pint to several quarts each.

After dinner a score or more of these cans were put on the train and away it went. As the little stations down the road were reached the milk was unloaded, a pint here and a quart there. At Hawthorne, which is the county

A MILK STATION—HAWTHORNE.

seat of Esmeralda County and the home of some of the most important of the railroad officials, ten or a dozen of these milk cans, including one that held several gallons, were put off the train. While the train tarried a young man sauntered up to the platform, chatted awhile there with the train men, turned over the wooden labels on two or three cans with the toe of his boot, found one with his name on it and carried the can away. A big black dog that all the habitues recognized and spoke to, smelled of various cans and eventually took the bail of one in his mouth and walked away with it. Each can was for a different family or for a hotel or a restaurant. The railroad company is the sole dealer in milk along its route, and takes this method of supplying customers who will take certain quantities regularly every day. People who do not want to buy milk in this way, get their supplies from the Croton Valley above New York City. It comes condensed in one-pound cans from the factories of the company, well-known there, and well-known throughout the Great American Desert as well.

Save for an oasis here and there, such as Wabuska and the Owens River Valley, the route of the Carson & Colorado Railroad is through a treeless desert. Sage-brush and grease-bushes are the only wood-producers to be seen on this desert and, as the reader knows, the trunks of these bushes will average about an inch in diameter at the most. One travels the length of this road without seeing a tree fit for fuel even on the oasis, and yet the locomotives burn wood for fuel. The explanation of this apparent paradox is found in the fact that many mountain tops along the route have a scattered growth of the nut-pine at an elevation of 9,000 or more feet above the sea level. It is this growth that supplies the locomotives.

One station on the road is on the Walker Lake Reservation of the Piute (or Pah Ute) Indians. These Indians ride free on the Carson & Colorado, but must ride on the freight cars only. At the Reservation station, my attention was called by a passenger to what was considered a novel spectacle. Three buck Indians were at work. They were carrying sacks of grain from a wagon to a freight car. They embraced the sacks as a country lad embraces his sweetheart at a corn-husking—by clasping them to their breasts—and so carried them, with many grunts and much perspiration, to the car. A bevy of Piute belles, gorgeous in calico apparel, and with faces striped and painted in barber-pole hues, looked on idly. Things are not as they used to be among the Piutes.

Quite as interesting a view of modern Piute life as this was had at Rhodes' Marsh, further along the road. Rhodes' Marsh, as will be told elsewhere, is one of the borate deposits of the desert. It belongs to the Nevada Salt and Borax Company. The borate is found here in little globules called cotton balls, nested in beds of clay. The proprietors employ about 125 Piutes to dig these cotton balls from the clay, and a regular Indian settlement has been formed near the railroad station. Bucks and squaws, old and young, work together in the clay, and in consequence, when pay-day comes the bucks draw considerable sums of coin, for the pay is good.

Once upon a time, though contrary to law, and against the policy of the borax people, these bucks exchanged coin for fire water, and got fighting drunk. A fight in that country, unless quickly quelled by sober heads, means murder, and the white men at the Marsh were obliged, in self-defense, to look after the drunken bucks. They did this by putting the cars that can always be found on the switch there, to a novel use. They made jails of the cars. Thereupon came a novel railroad wreck. Six bucks and two squaws, all half drunk, were incautiously locked in one car. After they had been in the car an hour or so they suddenly stopped yelling. A moment later the pine boards inclosing one side of the car burst out with a great crash, and the prisoners fell in a heap by the track. They had united in one grand assault on the car, with success. This was an exceptional case; as a rule the Indians are sober and industrious, and are earning their own living, white-man fashion.

Rhodes' Station is interesting in other ways. The main building there is a combined passenger waiting-room, freight depot, salt mill, and borax factory. The boilers and machinery used in making borax from the crude material, and in grinding for table use the salt also found on this marsh, are in the building used for railroad purposes.

There is a novelty about the locomotive on this road not

AN INDIAN WICKIUP.

readily noticed by the tourist. It is a spout, with a hand-valve, on one side of the tender. The tourist will surely see it at Rhodes', however, for, as the train rolls into the station a lot of Indians will be seen running in a throng from the wickiups, all carrying pails, or big square pans. Even before the engine stops they reach the spout, and trotting along beside it begin to draw water from the tender. A flowing well supplies water good for some purposes at Rhodes' Marsh, but it is not good to drink. The Indians want water to drink, and not for some purposes. The water on the locomotive is taken from very good springs.

Another interesting sight at Rhodes', though one in no way connected with the Carson & Colorado Railroad, is an almond-eyed Piute baby. A number of Chinamen are employed about the station. The whites expect that baby to develop into a very tough citizen.

Among the natural features of interest along the road is Walker's Lake. It is a picturesque body of water, full of the salts that make Rhodes', and Teels', and the Columbus marshes interesting. The geology sharps say that these marshes were once lakes like Walker's, but the sun and the sand got the best of them because they had no stream like Walker's River to back them in their struggle for existence. The lake is especially interesting to a tourist who knows sport when he sees it, for thousands of wild fowl may be found there in the migrating season. It has plenty of fish, too, including salmon of large size, it is said. A number of precipitous mountains rise along the west side, mountains that are full of gold and silver which the white man can not and the red man will not get, for they are in the Reservation. If the tourist is lucky he will pass this lake when a sudden squall comes out of the mountains. The way the wind pounces down on the water and whirls it aloft in towering eddies, hundreds of feet high, is a sight that alone counterbalances even the discomforts of a day's ride on the Carson & Colorado.

Moreover, the proximity of the lake to the track enables the train man and the old traveler to impose on the credulous tourist later on. After leaving Belleville, bound over the southern division (here the trains run every other day), the traveler skirts along the brim of a bowl-shaped valley, in the bottom of which lies Teels' Marsh. Summer or winter, unless the sunlight falls on it exactly at the right angle, this marsh looks precisely like a lake of water, such as Walker's Lake is, although its surface is one great stretch of white salts of various kinds. As the train rolls along, the conductor says to the tenderfoot:

"Pretty lake, that."

The tenderfoot admits that it is.

"Yes, sir," continues the conductor, "and it's deep enough for navigation, too. Lots of business done on that lake. See that steamer?"

Sure enough, there are two black smoke-stacks rising from a frame structure that looks like a Mississippi freight boat right on the edge of the lake. Black smoke is rising from the smoke-stacks, and good eyes can see a collection of houses near by on the shore. The tenderfoot is likely to see, too, smoke rising from single stacks scattered at intervals about the lake and to speak about them.

"Yep, they're little freight boats," says the conductor. "Mighty lively mining camps about that lake. That on this side is Teels'. Over beyond is Marietta. You would be astonished if you could drop over there, just now."

Then the conductor goes out and leans across the car-brake and snorts. The smoke-stacks belong to the borax works.

So far as traffic is concerned, the Carson & Colorado may be said to depend wholly on mining camps. It is a curious fact that as soon as this road had reached a camp in the course of its construction, a large part of the life and activity of the camp fled forever. The explanation of this, as told elsewhere, is in the fact that the freighters, with their teams, who had carried supplies to the camps, found their occupation gone, and leaving, took away a very lively and picturesque part of the community—the teamsters and their assistants.

There are very likely other interesting features of this

road, but I have only two left to mention. Both are remarkable.

The Carson & Colorado has throughout its entire length of 300 miles a population all told of less than six thousand people to draw traffic from.

The Carson & Colorado managers are able to pay running expenses.

CHAPTER XI.

TOLD OF A DESERT JOURNEY.

SHOULD any reader of these naked sketches desire to make a tour of this desert region for himself, he will find either Daggett or Mojave the nearest railroad station to the chief center of interest—Death Valley—but an objection to starting out from either of these places is the difficulty of obtaining suitable horses or mules. The Government expedition of 1891, under Dr. C. Hart Merriam, had to go to San Bernardino for an outfit, and except by a lucky chance any one wishing to follow their route would have to do the same. These near-by railroad stations are on the desert, and on a part but little less arid than the immediate vicinity of Death Valley.

A first-class outfit of horses and buckboards or wagons is needed. Not that the distance is so great (it is 166 miles from Daggett via the Amargosa Valley to the ranch in Death Valley, and 147 back by the way of Pilot Butte), but there is not a house along the way where any sort of supplies can be had; the springs are so far apart that water must be carried in kegs for the animals as well as the party; even hay, as well as grain for the animals, must be taken, for the bunch grass can not be depended upon, and last, but by no means least, the road is so rough in the washes that only the strongest vehicles and animals can stand it, and even then accidents impend at every turn. And an accident on the desert a hundred miles or more from any kind of a settlement is a very serious affair. The truth is, that such a journey as a tourist would wish to make, would be nigh impracticable, were it not for the ranch in Death Valley at which supplies for the horses can be obtained. A prospector's trip is entirely another matter, for the burros get fat on grease-brush and cactus, the desert men say, and it is really true that they do well on the scanty feed to be found about the springs.

A NOON-DAY LUNCH.

My own journey was made in December, and was therefore free from any danger from heat. Because the weather is commonly bright and cool one can, in that month, enjoy the journey very much, although the Indian summer balminess of the air in October makes that the ideal month for the trip. The outfit consisted of two buckboards, one drawn by a pair of mules, and the other by a pair of horses of the breed known as cayuse, or kyuse, throughout the

Rocky Mountain region. The buckboards were first-class—had been built for use on the desert—and so having had the luck to get 150 pounds of supplies sent ahead to the Amargosa Valley in a wagon bound to the mining camp of Montgomery, there was a reasonable hope of a pleasant journey when, on the morning of December 1, we drove out of Daggett, headed across the Mojave River bed.

The interest of the tourist is aroused from the first. There is the river to begin with. It has but a slight depression for a channel and no other sign of water about it than the stunted willows growing on the flats. As we drove across it, the animals kicked up clouds of choking dust instead of splashing water. Near by this dry ford, a gang of men worked on a new bridge that was suspended fifteen or twenty feet above the sand. It was a curious thing to see a bridge there, but the need for one is unquestioned, for floods come down the channel, sweeping everything but the deepest-set piles before them, and there is plenty of teaming across there from the mines up in the Calico Mountains, some six or seven miles from the stream.

More curious than the bridge, however, is the dam to be found four or five miles above. It is a subterranean dam. The Mojave, for the greater part of the year, flows through, instead of over, its sandy bed; and, although invisible, the current of water is considerable. A company has sunk a dam to bed-rock, intending to thus force the water to the surface and run it through ditches over the desert lands adjoining the river-bed and turn them into productive fields.

Beyond the river-bed the trail led off across a wide flat on which another interesting feature of the desert—a dry lake—was found. Its surface is covered with a cream-colored scale of dry mud that makes a beautiful road for light rigs, though the heavy teams from the mountains cut it up badly.

Seen from Daggett on a warm day, this dry lake appears as a most beautiful sheet of water, and, in consequence, the Daggett citizens sometimes beguile the tenderfoot into a fishing excursion in spite of the fact that even the Mojave River is then dry.

Off to the left is a narrow-gauge railroad belonging to a mining company. It runs from a quartz-mill with seventy-five stamps to the little camp called Calico, built around a mine of low-grade silver ore, that pays good dividends because cheaply worked.

But one who is looking for the picturesque turns very quickly from the river, the dry lake, and the railroad to look upon the beauties of the Calico Mountains. They bear a wretched name, but it is probably as accurately descriptive as anything the early explorers could think of. The name was given to them, undoubtedly, because of their colors. The foliage of green trees, with changing hues in autumn, has been denied to these ranges, but every peak, every face, every ledge, every declivity, every gorge, every strata, every rock, has a color of its own, and there are no two breadths of color exactly alike. They vary from marble white to lava black, from the palest green to the darkest carmine, from the faintest cream to royal purple—there is every tint and every brilliant and every dull body of color, and all mingled, contrasted, and blended, and all piled up in such magnificent masses as are beyond description.

The trail through the Calico Range is heavy with sand, and in places precipitately steep, but it had to be made in order to reach the borate deposit then known as colemanite. Since it cuts off several miles from the old route across the desert to the north, it is used by the prospectors and others, as well as by the borate teams.

Beyond the mountains lies Paradise Valley. This is a

misnomer or an exceedingly appropriate name, according as one looks at the valley. It is a grease-bush waste in fact, but seen in warm weather from a peak of the Calicos, with its surrounding rim of mountains, its buttes and hills, and its lakes apparently filled with sparkling water and

O'BRIEN'S GRAVE.

surrounded by groves of trees, it is, indeed, a paradise to look upon.

Beside the trail where it approaches the largest dry lake is a mound of earth, with a pine board at the head, on which has been written "John C. O'Brien." O'Brien was a prospector, who, in coming in from a trip on the desert, got out of water. He reached the coyote holes on the edge of the dry lake, but the water there only aggravated his thirst. He became insane, as the perishing always do, and began digging at last in the sand with his hands. His burros went into one of the Calico camps, but when help, which was delayed, did at last reach him, he was found dead under a grease-bush, with his fingers worn to the bone from his insane efforts to dig for water. The hole he had made in the sand was still there—a gruesome spectacle that set the driver talking about deaths on the desert.

Not a season passes that men do not perish, and not a man lives about the desert but has known of more than one such case. The effect of the heat is nearly always the same on the perishing, and the story of one is the story of all. They are found trudging along, carrying heavy burdens—dutch ovens, frying pans, worn-out mining boots and all sorts of useless camp outfit—while a burro walks beside with nothing on him, or is allowed to wander away entirely. Water has even been found on the animal where the man was dying for want of it. The dreams of water seem to become realities to the wanderer, and he believes he is at last in the water. More than one man has been found stripped naked and walking about on the burning sand, holding his clothing above his head. As rescuers approached, they were warned to be careful, for the water was deep, and once they were within reach of the dying man, he clutched them as a drowning man would do. The tourist will hear no end of such stories as this.

At the Coyote holes, a story of another sort was told to me. Some years ago, the superintendent of the silver mine in Calico used to pay off his men with coin, which he got at Daggett, and carried on horseback up to the mine. One pay-day, as he was going up with the gold, he saw one of his men coming down the trail, but thought nothing of that until the miner suddenly drew a six-shooter and captured both the coin and the animal the superintendent rode. The thief rode off around the east end of the Calicos, intending to strike off across the desert to Utah, but the stolen horse failed him near the big dry lake in Paradise

Valley. Thereat he turned loose the horse, walked to the Coyote holes, and lying down in the mud and water, worked himself wholly out of sight, save for his face, which was partly concealed under the grass that grew at one side of the spring. There he remained for hours, but an Indian guided a party of pursuers to the hole. The thief, seeing that he was caught, rose up and opened fire, but the others

THE "TIN LINING."

soon shot him to death, and then found the money in the mud under his body.

Thirty-one miles out from Daggett is Garlic Spring. The marks of old camps are found all about the spring, and of all other marks none is plainer than the tin can. The use of tinned goods is universal throughout the desert and the desert mountains. "Even the darkest cañon has a tin lining," as the driver said when he saw me photographing a heap of old cans in an out-of-the-way gulch of the Funerals.

Filling our water kegs at Garlic Spring, we drove on several miles and camped near the foot of Granite Ridge. It was an interesting camp for a first night out. The evening passed off delightfully. The sky was clear, the grease-bush fire cheery and novel to the tenderfoot, and the supper was made palatable by hearty appetites. At 3 o'clock the next morning we were routed out by a rain-storm, and since we had no tents, we hastened to roll up the bedding in canvas covers to keep it dry. Then we built a fire, and strove in sleepy fashion to get a breakfast. The tenderfoot had volunteered to make the bread, and did so; but, somehow, while trying to keep the rain out of the grub-box, he spilled an unknown quantity of salt in the flour he was about to mix. The drizzle of rain was so disagreeable that we were not inclined to cook any meat save ham, and this was almost as salty as the bread proved to be. We didn't eat very much breakfast.

We were joined at daylight by two unbidden guests, who, in silence, sat near the camp, watching every movement, and occasionally shifted from rock to rock. As soon as the teams started, they pounced on the scraps of salty ham and saltier bread by the fire, and ate it ravenously. These guests were great black crows. I doubted not that they would find food in quantity and quality by that fire to last them indefinitely—to kill them, in fact, if they ate it all—but within an hour a shadow sweeping along from the rear made us all look around, and there were the two crows, black and glossy, bound to Death Valley, if we were—bound at least to stay with our outfit as long as it was in the desert. The driver said that no team ever crossed the desert without silent black followers, fit to make a man's flesh creep—especially when the sun is hot and his canteen empty.

Among the rocks on the crest of Granite Ridge, the driver said:

"Right over there, some place, is a grave we can not see from the trail. In it is buried a man who kept a saloon at Daggett. He had been on a spree for a week or two, and came out here to stop with a man living at Cave Springs, a few miles beyond here, in order to sober up. The man at the Springs was not exactly right in his head, but he had

THE CAMP FIRE.

wit enough to file on Cave Springs, and sell water to all who came along, at two bits per man and per head of animals, for the night. The saloon-keeper carried a knife, saying to a friend, before starting, that if the Cave Spring man 'gets any of his weird streaks on I'll have something to do with him.' It was the saloon-keeper who got the queer streaks, however. The sun proved too much for the rum-soaked brain, and going up Granite Ridge he talked so much about lynchers being after him, waving his knife aloft the while, that the other man got frightened, and finally said:

"'There they come; you'd better jump out and run for it.'

"At that the saloon-keeper did jump out and run, kicking off his shoes directly, that he might run the faster. Nor did he stop running till he fell and died. Some prospectors found him by following his bloody trail.

"'All the flesh was cut from the bottoms of his feet by the sharp rocks he had run over,' said one who saw him."

We made a brief stop at the old ruined hut by Cave Springs to rest and water the animals, and then pushed on into the valley of Amargosa, though not the part of it where the old borax works may be found, that were farther on. It was getting late, too, and the rain-storm of the early morning, which had cleared away during the day, came on again (or another one came), cold and raw. We could see the snow falling on the mountains about the lower end of Death Valley and on both sides of the Amargosa, but we kept on past the Saratoga Springs (one of the most desolate places in the desert), and at dark began to climb a sandy wash or bed of a torrent, twelve miles long, leading to a crest in a bend of the Amargosa. Just nine miles beyond that crest were the buildings of the old borax works, and there we hoped to find comfortable shelter and a square meal. The horses were already tired when the climb began, while the trip over this divide was, by itself, a day's journey for them, but the driver said they could better afford to make it than we could to camp in the rain without shelter, and so up we toiled, those walking who could. Being unincumbered, I walked on ahead and alone. It was a most lonesome experience, but if lonesome to a tourist what must it not be to an unfortunate man brought there alone and by distress or accident.

Leander Lee (called Cub by the Arabs), the watchman at the borax works, was in bed when we arrived, but he and his wife (a round-faced, pleasant-looking squaw), and his daughter, a shy little lass of thirteen, turned out and went to the kitchen, leaving us in possession of the main room, where a big fire of mesquite wood blazed in a stone fire-place. We thawed out quickly before that fire and then went to the kitchen, where hot coffee with milk and sugar, hot flaky biscuits, hot mealy potatoes, and hot juicy ham were ready for us.

The most interesting features of this place are the so-called boiling springs and the immense clay buttes. The boiling springs are but bubbling springs, and one can safely put his hand into any of them. In the days when borax was made here the workmen constructed a rude shed so that the water of one spring ran through a vat convenient for bathing. They are said to have derived great benefit from these baths—not so much because the baths were medicinal; they were baths.

The buttes of clay stand like islands about the sides of the well-defined channel of the stream. They are water-worn on all sides, and the banks of the stream are composed of the same material, deeply eroded into capes and promontories. Along the center of the channel are shallow water-ways, and in places there are little pools of alkali water, while the ground is nearly everywhere moist, but covered with a thin, dry crust of carbonate of soda, mixed with dust from the sand-storms. But no water ever flows down those mud-channel water-ways for more than a few hours at a time, though torrents several feet deep come rushing down the gullies from the mountains, when clouds burst about the peaks. One Government explorer, who was in the region twenty years or more ago, reported the Amargosa as navigable, but neither "Cub" Lee, nor any other desert man has ever seen enough water in the channel to float a Whitehall boat.

From the old Amargosa Works to the ranch in Death Valley the distance is sixty-one miles, with no water along the road until Bitter Springs is reached, twelve miles from the ranch. However, there are two different springs—one at an old mine called the Ibex, and another some miles

THE DESERT SIGN OF WATER.

beyond—which lies a few miles from the trail and could be reached in case of necessity, though to reach either would prolong the journey by a day. We therefore carried water for the one dry camp we were to make. The camp was a bleak one, just below the crest, under heaps of bedding that night, but Lee curled down in the shelter of two bales of hay and slept comfortably with but one blanket—at least he said he was comfortable.

The trail from the crest runs down a branch of the Furnace Creek Cañon into Death Valley, and here I first saw the desert men's sign of water. They have the fashion of building up monuments of broken stone on prominent points about gulches where a spring can be found. Then at the mouth of that particular gulley where the spring is, they drive some sort of a stake and put up an old tin can over the top of the stake. The Indian water-sign is a white stone put on top of a monument, but the only sign a man with a shovel and strength to use it needs, is the common mesquite tree. If he can dig below mesquite roots, he will find water and usually good water. So says "Cub" Lee. In some parts of the desert, the springs may be located by white deposits on the rocks round about them—good springs, at that; but there are others, like Granite Spring, where there is neither tree nor grass blade to mark the flow, while one sweet spring in the Funeral Mountains rises up within three feet of a salt-water brook, and there are salt springs around it.

We camped at Bitter Springs, because we wanted to visit some mountains of borates next day, of which something will be said elsewhere. The next day was therefore passed in the Funeral Mountains, among the gulches leading into the Furnace Creek Cañon. The tourist will here find broken lands of the most picturesque forms. Save as they form the eastern wall of Death Valley, the Funeral Mountains have been most outrageously named. They are as gay-colored mountains as the eye ever saw. One ought not to be too critical in this matter, however; indeed, a discriminating tourist will rather feel a sense of relief because no pious crank has been found among the explorers to name its principal peak Horeb or Pisgah, as was done in the wild region about Asheville, N. C., and elsewhere on the continent. The lofty and many-colored peak that overhangs the northern side of the Furnace Creek Cañon is called Naghi, and that in the Indian tongue means Sheep Mountain. It is the home of many mountain-sheep. On the south side lies Monte Blanco, and that is descriptive, too,

FUNERAL MOUNTAINS.

for its face is white with the deposits of borates found there.

Under the shadow of Monte Blanco we found a countless number of clay and soda peaks, but once we had passed through these, we got into steep-walled gulches, some so narrow and deep that crevice were a better name for them. Here the rocks have been torn and split, and even twisted and warped. Strata of rock formed by the slow water deposits during ages lie next to strata that came in molten floods from ancient volcanoes, and, when these had cooled, the earthquake came and tossed them up into such massive cliffs and rugged peaks as amaze and awe the spectator. No vestige of shrub or plant can be found in these gulches —the picture is one of Nature in her naked beauty.

Furnace Creek rises in a host of springs well down toward the lower end of the cañon. It is marked, from its source until it spreads out fan-shaped at the mouth of the cañon, with an abundant growth of mesquite, the Indian arrow-reed, and other plants. The water is warm and slightly alkaline. It has been estimated to run from 100 to 200 miner's inches of water, but if one may judge by the proportion of it used in irrigating the thirty-acre ranch down in the valley, it would irrigate at least 160 more.

What we saw in Death Valley, save one thing only, will be told elsewhere. On the journey out of the valley, we passed five graves in one day, of which but one held the body of a man whose name was known—that of Bill Shadley. The rest were graves of unknown men, who had been overcome by the heat while traveling the regular trail, and, dying where they fell, had been buried by strangers. These were the graves of men who stuck to the trail even in their delirium. If five bodies lie so, how many have wandered away to die where their bodies, though preserved by the action of the salts of the valley for months, were beyond the sight of the wayfarer on the trail? Not one of the party but constantly felt during our journey that the body of some unfortunate might be found near the trail at any turn.

We camped the first night on the trip out of Death Valley by Mesquite Well, near the lower end of the valley. It was a pleasing desert camp, being well sheltered by the mesquite grove, which also furnished abundant fuel. The water of the well is counted excellent, but it had its drawbacks for the unaccustomed traveler. We had to skim out a number of drowned long-tailed rats and boil it before using it. No doubt it was then healthful if not pleasing. Because it illustrates life on the desert, let it be further said, that at Granite Spring the water was worse. The spring is an unprotected hole in the hill-side, and some one had, during the past year, kept a small bunch of cattle feeding on bunch-grass in the vicinity. These cattle had crowded

CAMP AT MESQUITE WELLS.

about the spring until nothing but absolute necessity made it usable. In fact, we found nothing but two pleasing springs in the journey of 369 miles by buckboard. One was Hidden Springs, 3,300 feet above the level of the sea, on one of the Panamints west of Death Valley. This we reached the second day from Death Valley Ranch. The other was at an artificial oasis made by Mr. John W. Searles in a cleft in the side of a granite peak in the Argus Range, 7½ miles from the San Bernardino Company's Borax Works.

The emergencies likely to arise in a desert journey were illustrated on our way out of Death Valley. As we planned the journey, we camped at Mesquite Well the first night. The second night we should have camped at Hidden Springs, and the third night have reached the San Bernardino Mining Company's Works, which lie twenty-five miles off the usual trail to Daggett. But, on the second day, it was found that one of the horses was getting used up, while one of the mules began to bleed at the nose. We camped nine miles short of Hidden Springs the second night, and the animals had not half a drink in the morning.

At Hidden Springs they got all they wanted, and a bite of the long green grass that grows there as well, but we found before long that it was useless to try to reach the works that night. The supply of feed had been ample for the journey that should have been made, but, as it was, the animals had less than half rations of hay that night. Of course, it was a dry camp. Besides, we had a drive of thirty miles or more to the works on the fourth day. This was "a tight squeak," as the driver said, but we made it just at nightfall, and our trouble was over.

As told elsewhere, the works are under the supervision of Mr. John W. Searles, famous in the history of the State as a bear-hunter and prospector. It follows naturally that we received a hearty welcome. He literally killed the fatted calf—drove a fat steer into a shed and butchered it to supply the table with fresh beef. A man of extensive knowledge and wealth, Mr. Searles might almost be called an Arab, for he makes his home on the desert, but his home is different from any to be found in all of the arid region of the country.

We got a hint of what it was like in the abundance and variety of the fruits on the table. The hardy Baldwin apple reposed beside the tropical fig and orange, not to mention a dozen other kinds of fruit, and all "from the ranch."

The second day of our stay, we drove up to the ranch behind a vigorous team, that made light of the load and the difficulties of the trail. And such a trail! It is a twisted gash in the side of a precipitous granite mountain. At the top of this gash, a number of springs flow freely. Taking advantage of these, and a number of little narrow benches just below them, Mr. Searles has built a terraced garden. No part of it is more than fifty feet wide and no level 200 feet long, and there are not a half dozen levels. The soil is decomposed granite mixed with manure brought from the works. An elderly Chinese servant is kept in a little house here, and on these narrow holdings Mr. Searles produces every year more fruit than the forty-odd men about the works can eat. In its cooling shade, its flowers, its fragrance, and the utterly barren rocks that rise precipitously on every side, it is an ideal oasis.

A striking feature of the scenery about the old lake-bed, where the borax is found, is a host of stone "needles" that rise at the south end. They are picturesque pinnacles of a brown rock-like stalagmite. A similar formation has given the name of "The Needles" to a station on the Southern Pacific road.

After a too brief stay at this place, we headed away for Daggett. On the first day we drove about thirty-five miles to Granite Spring, where we camped beside the grave of the only horse-thief I heard of in my journey over the desert. The grave is a rude pile of rocks, and some one has erected a head-board, on which were carved the words: "Grave of Gus Meyer." With two Spaniards, Meyer tried to run a bunch of horses from the Los Angeles country across the desert to Utah. The outfit got as far as Granite Spring, and camped there one night. Then came the owners of the horses, and when Meyer walked from a little stone hut that stood by the spring, then to a point of rocks near by, he was shot dead. The Spaniards fled in the darkness and escaped into Nevada, where, at a mining camp, they told a story of the shooting that made the miners suspect they had killed the man. So an investigation followed, the body of the thief was found and buried, and the facts determined. This done, curiously enough, the Spaniards, who had been arrested, were let go.

But two incidents remain to be told. On the last day's journey our route lay through a most picturesque grove of the Yucca palm. Seen from a distance it is as if some one had planted an orchard, thousands of acres in extent, and then left it uncared for, so that, while many trees thrived, others drooped and some died, leaving vacant spaces. Seen near at hand, the Yucca is a curious tree, little like anything in an orchard, for, as the driver said of it, "it is a tree with its fingers all thumbs."

The grove was the brightest picture of a long, and, because we were impatient to reach a mail once more, a wearisome day. We arrived at the station just in time. In spite of good care, one of the mules died two days later. Had we been obliged to drive it another day, it would have died in the harness, and some of the party would have walked in.

CHAPTER XII.

ABANDONED DESERT MINING CAM S.

ONE feature of the desert, which no tourist could overlook, is the abandoned mining camp. I visited three of these camps in the course of my journey, and curious tales of life in several other camps were told to me. Of the whole number, none was more interesting than Marietta, on the side of the Excelsior Range, northwest of Teels' Marsh, in Esmeralda County, Nev. It is nine or ten miles back from Belleville, on the Carson & Colorado Railroad, the wagon-road to it leading over a steep pass in the mountains, of which the peak that rises to the right of the trail is 10,250 feet above the sea, and the pass itself over two-thirds as high.

It is a ghostly experience to visit such a camp as Marietta. The place is seen, four miles away, from the road that winds down the cañon in the valley. The old red quartz-mill with its tall black smoke-stack, the rows of wooden and adobe houses, even the little rock huts which the miners built for bachelor halls, stand out in the pure air so clear-cut and distinct that they seem but a step away. The mountains, with their copper-stains of green, rise in rugged beauty beyond. Everything seems so complete and in such good order that the mind can not resist the impression that a thriving community lives there. Scarce does this impression fade away on nearer approach, in spite of certain tumble-down adobe walls, for numbers of the houses stand with doors closed and glazed windows intact, the handle of the town-pump still projects invitingly, while the wooden walls of some of the houses show only trifling weather stains. But once the traveler gets within the limits of the settlement, the smokeless chimneys, the vacant homes, the empty shelves in the stores, and the utter silence—it is as though one had unexpectedly found himself in the midst of a collection of skeletons and graves. There are, indeed,

VIEW OF MARIETTA, NEV.

graves to be seen close at hand, while a closer inspection of the buildings shows traces of plenty of deeds of blood—an awning post pierced and splintered by a heavy bullet, a hole in an adobe wall where another bullet had entered, traces of a splash of lead on a stone wall, where still another projectile had flattened. Nor is the feeling that this is but the ghost of a town much dispelled by the coming of a white-haired, white-bearded old man, in faded attire, from one of the smaller wooden buildings, the sole inhabitant of the town, the man who is left as watchman over the mill and mining machinery.

But, if ghostly now, there was a degree of life there in the year 1880 that would have startled a peace-loving tourist as much as its weird aspects awe him now. Then 163 men were working in the principal mine, and the roar of

AN OLD MAN IN FADED ATTIRE.

the stamps, where thirteen men worked in each shift, never ceased. There were scores of small-claim workers and prospectors. There were thirteen saloons, in some of which day and night shifts of dealers played faro and other games of chance. There were five houses run by sporting women. There were three stables with horse-yards attached, one of which would accommodate 200 horses. One supply-store took in from $13,000 to $17,000 per month and another but little less. A daily stage with six horses before it thundered down the pass in the Excelsior Mountains, bringing mail and eager seekers for fortune—a stage that was held up by road-agents four times in one week; was robbed thirty times in all. The town of Marietta was as full of mountain life as a Furnace Creek den is of rattlesnakes, and no one can sketch the pictures of the scenes there with more graphic pencil than Thomas Purcell, the white-haired watchman, who is now all that remains from its stirring career.

"Those bullet marks? They were made in the fight between Tom McLaughlin and John Brophy," he said, when I spoke to him about the signs of fighting. "That was one of the greatest fights the mining camp ever saw. McLaughlin and Brophy were good friends, too. Mac was working twenty-five men in a mine, and running the saloon with two shifts of faro-dealers over there, where the awning-post is shot through, and he bought two head of beef from Brophy, who was a butcher, every week. But they both had women, and the women quarreled first. Both wanted to be the leader of society, I guess, and the men got mixed up in it right away, and agreed to fight it out. And each side knew the other was game, and so they called on their friends to help them, and you bet the call was answered.

"That night Brophy's party—four of them—slept here at Mrs. Sperry's, and McLaughlin and his three friends over there next to the saloon. Everybody knew it was to come, and the women and children were hustled off there among the rocks out of range, except John Brophy's. He took his down to his slaughter-house, and while he was there his friends, led by his brother Hank, opened the fight. They'd got their breakfasts and were walking down there by the pump—Hank Brophy, Dick Gillespie, and Hank Hankins—waiting for Tom McLaughlin to come out.

"They didn't have long to wait. Tom only waited for a bit of a smoke after breakfast, and then, after laying plans with Tom Taylor, and George Martin, and Fred Schofield, he walked out of the front door of his saloon, revolver in hand.

"With that, Hank Brophy opened fire and the rest joined in. The roar of the exploding cartridges was continuous —there wasn't any yelling; they were too much in earnest for that—but most of them kept jumping about to kill the aim of the other side, while McLaughlin walked straight down toward Hank Brophy. Tom Taylor was laying for Hank, too, and just about that moment John Brophy saw that they had a cross-fire on Hank, and started to run to help him. John's wife grabbed him, but he flung her off, and with a rifle ran up the street and drew a bead on McLaughlin. His finger was on the trigger when Taylor fired at him from behind that 'dobe wall over there. The bullet went through him just below his heart, his finger-nerves contracted and pulled off the rifle, the bullet struck that stone wall there, and over he dropped dead, the first one in the fight. An instant later McLaughlin shot Hank Brophy in the shoulder, and then fell dead himself—Dick Gillespie did it, I guess. Then Schofield got Dick, while it was Hank Hankins, may be, who dropped Tom Taylor. They were all shooting so fast that no one exactly knew how it was, but four of them were killed. Gillespie and John Brophy lie over there in those graves—those heaps of rock where the little fence is—but Tom McLaughlin and Tom Taylor were Odd Fellows, and were carried over to Belleville for burial.

"What became of Hank Brophy's wound? He got well fast enough, and went to Arizona. There he got into trouble about some cattle, and when Tom McLaughlin's friends here heard he was in custody they went over there and hanged him.

"Then there was the case of Corbett and Rogers. They went broke in Columbus, which was a lively camp, too, in those days, and asked a man driving up to Candelaria for a ride, and got it. In the hills they murdered and robbed him, and eventually came along the road here in Marietta afoot, with the sheriff only a few miles behind them, having three well-armed men in a carriage. The two got something to eat here, and went on up the pass toward Carson. Pretty soon comes the sheriff's posse, and they stopped here for a drink. Curious about that drink. The time it took to get the drink gave Corbett and Rogers time to get off a soft piece of road and well on to a rocky one. So when the sheriff drove up behind them they heard him rattling over the stones before he saw them, and hid behind the rocks, so that he drove past unsuspectingly. But they did have nerve! They followed right up, and, seeing from a curve in the road where the sheriff's outfit had stopped at the stage-station, and had gone in for a drink, leaving one man to guard all the guns and horses, they slipped up quietly, held up the guard, got into the buggy and drove off with all the guns and a pretty good snack of grub. The sheriff had to walk back here and borrow money to get back to Candelaria. They joshed him terribly over it.

"Corbett and Rogers drove off by the way of Walker Lake, where they represented themselves as the sheriff's posse, and so were able to get supplies, and eventually got to Eureka, having robbed a freighter of two good horses en route. Word of this reached the sheriff in Eureka while they were there in a stable, holding up the hostler, and stealing two more good horses. Here they lost their luck. They got started out of town all right, and there was a lively race along the trail, but the sheriff overhauled them, killed Corbett and wounded Rogers. Rogers got well and was sent to the pen for ten years."

"What, ten years for murdering a man for money?"

'That's all."

Purcell is about as interesting as his stories. He is a native of Flushing, Long Island, but he has been in the mines nearly all the time since the first rush for the Pacific Coast. He has had plenty of good claims, has had plenty

PANNING IT OUT.

of money, and has now got, as the result of a life's work, a pension of $10 a month because of service during the war of the Rebellion, his salary as watchman, and some more claims. I had never seen free-milling gold ore worked, and he was pleased to show me the prospector's plan.

Three lumps of a brown rock, the size of a walnut, that showed under a glass dull yellowish specks, were put in an iron pot, made of the end of a mercury flask. Stooping down, he pounded the rock to powder with the head of an old drill. Then he got his pan—a shallow pressed iron pan, rather smaller and more flaring than a milk-pan—and put the dust into it. Pouring on a quantity of water, he swished the mixture back and forth with a peculiar twitch that slopped the sand and water in thin sheets over the edge until at last but a teaspoonful of the mixture was left. Then, with another twitch, he slumped the bulk of the sand to one side, leaving a thin tail behind—a tail of pure gold. The sight of the yellow dust was fascinating. To have seen it washed out is to understand how it is that once a man is a miner he can rarely become anything else.

Belleville, on the Carson & Colorado Railroad, is another village of the sort, though there is a trace of life there now, because it is the end of the south division of the railroad. To tales of bloodshed there must be added one of a lynching, and it is a story that illustrates still further the phases of desert mining camp life.

The murder was simply one of common brutality. Charles Marshall was employed in the Belleville livery stable, but was discharged for drunkenness. That night he entered the stable, where he found the man who had been hired in his place, and saying, "You 've got my place, but I 've come to fix you," pulled a revolver and shot the fellow down. Then he kicked him as long as there was a sign of life, and went away.

"There's the old stable over there," said Tom Pepper, a miner who was in Belleville at the time, "and here was the hotel where the inquest was held. There wasn't any doubts about the facts of the crime, but they nearly hanged the wrong man; there was a witness to the crime and the facts were brought out the next evening. Marshall was kept in a room back of the hotel, and there was a barber-shop right beside it. About midnight the vigilance committee entered the hotel, carrying a rope, and a handful of cotton waste, with a thin strip of canvas for a gag. The constable guarding Marshall was some distance away, and so in some way the committee got into the barber-shop instead of Marshall's room. The leading doctor of the town slept in the barber's chair every night, and, as he was about the build of Marshall, the committee grabbed him in the dark. They were putting the gag to his face when he awoke, realized the situation, and managed to gasp just in time: 'I 'm the doctor; I 'm the doctor.'

"The committee backed out to meet the constable who was in charge of the prisoner. He was so badly frightened at the sight of the masked men with a rope, that he was unable to put the key in the door to unlock it at their command, but one of the committee soon unlocked it, and all filed in. Marshall was sitting in a chair paralyzed. They quickly surrounded him, and one grabbed him by the hair. At that he groaned, 'Oh, my God, boys, don't!' but the words were all but smothered by the gag of waste that was tied on. Then he was jerked to his feet, led and dragged across the street, a rope was thrown over the crane there, and in a minute more he was swinging in the air.

"I was working on morning shift in the mill then, and came on at 4 o'clock. Going to work, I had to pass right by the crane. I 'd noticed Marshall had a pretty good pair of boots on, of about my size, and I thought I 'd get them as I went along, but when I swung his body round, so that I could get hold of a boot, I happened to look up, and there were his eyes just bulging out and staring down at me. Wow! I didn't want any boots after that."

As we drove into Belleville, Tom pointed out a place where a Chinaman had been killed.

"White man do it?" said I.

"No, Piute."

'A fight?"

"No—o—o, no fight. Just shot him to see him drop. Nobody ever cared about a killed Chinaman, you know, nor for a killed Piute, for that matter. Why, above here, at Rhodes Marsh, a Piute girl died in childbirth. That was such an unusual thing that the others thought she was bewitched. Her father was very old and blind, but he took a knife in his hand, got a boy to lead him to the tepee of

AN ABANDONED MILL.

the old squaw, who they thought was the witch, and there, although her husband, a grown son, and two daughters were in the tepee, the old man cut the woman all to pieces. Nobody ever did anything about it except to bury the old woman."

Columbus, not far away, is another abandoned mining camp, but the stories of the murders and the lynchings there are in no way distinctive.

Down in the Death Valley part of the great desert are other mining camps, now idle, and likely to remain so indefinitely, unless a railroad is built across the country, for water is scarce, and fuel, suitable for a quartz-mill, scarcer.

Resting Springs, over in the Amargosa Valley, east of Death Valley, has a ten-stamp mill. The old Ibex mine stands near the road, leading from the idle Amargosa borax works to the idle borax works in Death Valley. There is a five-stamp mill there owned by a Chicago company. "It has been running occasionally during the past two years," says the report of the State Mineralogist for 1889. "That it is not run constantly is owing to lack of fuel."

In 1873, the mining district of Panamint was organized, because of a vein found in Surprise Cañon, at an altitude of 6,600 feet. The district includes part of Death Valley. The camp of Panamint was built at the mine, and a great quartz-mill, with an engine of 120 horse-power, was built there by Senators Jones and Stewart. There was a tremendous excitement on the coast over this desert find, and the road thither was thronged with an eager, reckless crowd. Many went afoot with wheel-barrows, or harnessed to buck-boards, because they could not buy horses to haul their supplies. Then the mill burned down and away went the campers—some perishing on the desert for want of water. Out of the debris a ten-stamp mill was made, but even that is idle for lack of fuel, and only one white man remains. A more curious character than he is said to be would be hard to find. The white Arabs speak of him with much respect as one who writes novels and other literary matter, and has sufficient influence to keep an open mail route to the deserted camp.

A handful of Indians who call themselves a distinct race —the Panamints—live about the old camp.

Although nothing but borax has been made commercially

available within Death Valley itself, there is, according to the mineralogist report just mentioned, a deal of good ore in the regions round about—fairly good ore only. If on a railroad, that could bring cheap fuel, it would make many men rich; but, lying in the desert, it has been but a mirage to lure men on to bankruptcy.

CHAPTER XIII.

A CALIFORNIA BEAR HUNTER.

PROMINENT among the names of the gold hunters of California stands that of John W. Searles. It is prominent, not alone because he made a success of his prospecting, but because he was one of the most experienced hunters of the grizzly bear the State has seen. It has been some years—more than a score—since Mr. Searles hunted the king member of the bear family, for he has been too busily engaged in the borax business to spend the time in the mountains. Moreover, his last encounter with one of the tribe was of a nature to cool even the ardor of such a veteran as he.

I had heard that Mr. Searles was the hero of a terrible bear fight, and so, when I reached the San Bernardino Borax Mining Company's works, of which he is superintendent and chief owner, I took the first occasion to ask him about it. He smiled through his bushy beard and eyebrows, and, turning to a desk, took a two-ounce bottle from a drawer and held it up. There were twenty-one pieces of broken bones and teeth in the bottle. Then he took an old Spencer rifle from a corner of the office, and passed that to me. There were not only a number of dents in the stock, but one plainly noticeable in the top of the barrel. The bones and teeth in the bottle had been crushed from the lower jaw of Mr. Searles by the bite of a grizzly bear, while the dents in the rifle were made, in the same fight, by the grizzly's teeth also.

I was asked to put my hands among the dark-brown whiskers, just beginning to turn gray, on the chin of Mr. Searles. There were dents in the jawbone on both sides that seemed to half cut it off. At about this time I noticed, too, that Mr. Searles could not readily turn his head. He had plainly been pretty well chewed up.

It was on the 15th of March, 1870, in the mountains of Kern County, Cal. Some time before that, Mr. Searles, with others, had gone off from the settlement of Visalis for a month of sport with the deer in the mountains. They were in a part of the country neither had visited, and so had taken a guide along, who professed to know the haunts of the game, but, for some reason, they did not have much luck at first. However, Mr. Searles eventually saw a big buck upon a ledge, and, getting a shot, knocked it over a precipice as it ran, and thus scored the first kill.

J. W. SEARLES,
The California Bear Hunter.

Going to the edge of the precipice, to look over and see where the game had fallen, Mr. Searles saw two full-grown grizzly bears and a cub half-grown in the cañon below. Thereat he managed to get around close to the animals, piled one of them dead across the dead body of its mate, and, as the third fled down a precipitous trail, knocked it end over end, hand-spring fashion, with a bullet in the base of its head. It was a remarkable bag of game that sent Mr. Searles into the fight that so nearly took his life.

If Mr. Searles wanted grizzlies, why, another part of the mountains was the place, so the guide said. There were two there that had been killing cattle for a long time, and they were not only large, but bold and ferocious.

That was the kind of bears that Searles was looking for in those days, and away the outfit went. They reached the spot and pitched camp, but, because of foul weather, did not see the grizzlies, nor have any fun to speak of. Meantime, Searles had shot away about all of his cartridges, and sent for more, after the fashion of those days, by hanging his order on a bush beside the stage road. The order included an empty cartridge-box, but the stage-driver threw the box away and then got the wrong cartridges. Searles found, however, that he could hammer the cartridges through the lock after trimming the bullet carefully, though it took two blows of the hammer to fire a cartridge when in the barrel. So he kept on hunting.

Then came a day, when, with four of the right or old cartridges in the magazine, and the rest of the number of the whittled kind, he started out on horseback, although the brush everywhere was covered with snow. Four miles from the camp he tied his horse, and then went poking about afoot. So it happened, that, as he walked along the side of a gulch, he saw through the brush a big grizzly lying in a bed. He could see no more than its nose, but aiming low he let drive and rolled the brute over, when two more bullets finished it.

Working his way down, Searles cut the beast's throat, and stood beside it, pressing with a foot on its breast to make the blood flow, when a noise was heard in the thicket hard by. Nothing could be seen, but Searles knew the sound, and after a time found the trail of another bear.

By this time the afternoon was wearing away, and Searles was wet to the skin from the moist snow that covered the brush, but he took after the bear with all the ardor of youth. He eventually located the beast in a chaparral thicket, but worked about it for some time before getting a sight, and then, all at once, to the very great surprise of the hunter, the bear rose up on its hind legs, with its nose not eighteen inches away. It was impossible, because of brush, for Searles to back off even a step; the best he could do was to point the rifle across his body as near as he could guess toward the base of the beast's jaw, and pull the trigger, hoping to send a ball into its brain. As the gun was discharged, the bear pitched over on his fore-feet, gasping and pawing at its eyes, where the flame of the cartridge had burned the hair, but apparently only a little hurt.

As quick as thought, Searles threw a new cartridge into the barrel, raised the rifle, and, pointing at the base of the bear's brain, pulled the trigger. It was one of the whittled cartridges, and was not set home. With another wrench on the lever, Searles tried again and failed. A third time he strove in vain to fire the gun, and then the beast turned on him, open-jawed. Searles jammed his rifle into its jaws, but it brushed the weapon aside, threw him to the ground, and with one foot on his breast bit him in the lower jaw. The next bite was in the throat, severing the wind-pipe and laying bare the artery, as well as the jugular vein, and then it grabbed the flesh of the shoulder, laying bare the bones and cutting a blood-vessel, from which the blood spurted up, so that Searles, lying there, saw it stream in a curve above his face.

As the bear pulled this mouthful of flesh clear, its foot slipped, and Searles rolled over. His coat was all in a hump on his back, and the bear bit into that once, and then went away.

"What does a man think when a bear is tearing him to pieces?" was asked, as Mr. Searles paused in his narrative.

"Twenty years in California, to be killed at last by a —of — of a grizzly, is what I thought. I remember lying there and thinking so very well. I was disgusted."

He was as near dead as a live man was, but a part of his discomfort saved him. It was turning extremely cold, and the wet clothing began to freeze, but this temperature sealed up the torn blood-vessels. Then, in spite of his horrible condition—with his lower jaw dangling about, and his throat in shreds, and his left arm useless—in spite of the most frightful pain, Mr. Searles managed to walk and crawl to his horse, to mount it, though it was a fractious beast, to ride to camp, and to reach Los Angeles hospital, a three days' journey away. He lived while surgeons consulted over the best way to make him comfortable during the short time he had to live, and while they talked about boring through sound upper teeth, in order that they might wire the pieces of the lower jaw together and to the upper one, he even managed to kick one of them from the bedside. Then one came who patched, and pieced, and sewed, and plastered, and inspired hope, and in three weeks the old hunter was up and around, getting well in a way to astonish even the surgeon who had pulled him together.

As early as 1862, John W. and Dennis Searles, brothers, had prospected in the Slate Range of mountains, the second range west of Death Valley, and with success. They had worked arrastras and little mills there, Los Angeles, more than 200 miles away, being their usual place of departure. Their camp in the Slate Range looked down on a wide marsh, that glared in the hot summer sun like a pool of molten silver. The marsh was supposed to be a vast bed of salt and carbonate of soda. The carbonate of soda was used in working the ore in the mines, and an engineer they

SAN BERNARDINO BORAX MINING CO.'S WORKS.

employed used to complain that the stuff had borax in it, which interfered with its proper influence on the ore. Some time in 1863 or 1864, John Searles heard about the borax in Clear Lake, which a company in San Francisco was exploiting. So Searles took certain samples of incrustations from the marsh below his camp, and brought them to San Francisco, and showed them at the office of the borax company. He was very soon invited to go into a private room. There he found several wealthy capitalists very much excited. There was a discussion of the matter, and then an agreement was made orally to form a company, on terms favorable to Searles. This done, he was told to go back and get more samples and a greater variety. He says he did so and brought mud, water, crusts, carbonate of soda, and, last of all, some tincal, the naturally-formed borax. Mr. Searles did not know what it was then, but he knows more now and he remembers the crystals very well.

His samples very plainly excited the crowd, as before. Searles was invited to come again, when the samples had been analyzed. He came, and got a frigid reception. He was told that there was not a single trace of borax in any of his samples, and he was advised to prospect for copper —copper was the great metal to make a man rich.

Meantime, Searles had spent some hundreds of dollars at the behest of these men, and was at the bottom of his pocket. He said that as he had worked for them they ought to share the expense, but, if they would give him the cost of a passage to Los Angeles, he would call it square. The capitalists could not afford to put up a matter of $25, he says, and Searles went out on the street, met a friend, borrowed the money, and went away.

So it was not until 1873 that Mr. Searles had very much more to do with borax. He had put in the time pretty well, with gold and grizzlies, however, and was ready to take hold of the matter independently when the time came. He had, moreover, made some borax without knowing it. While working a gold mine in the Slate Range, he got some solutions from the marsh, and set them to crystallize in old powder-cans, buckets, and kegs. Thus he got a lot of crystals of carbonate of soda, supposing it to be borax, and carried it to Los Angeles, leaving a half-inch deposit of good borax in a keg, supposing it to be useless.

At last came news of the borax finds of F. M. Smith and others in Nevada, and it made a furor. With the news came a man named McGillivray, from a Nevada marsh, bringing a sample of borax as it was found there. Searles got a good look at this, and then packed an outfit for a journey to the marsh west of the Slate Range. His brother Dennis, Mr. E. W. Skillings (who, by the way, had been of the party when the bear fight occurred), and J. D. Creigh were in with John, and claims of 160 acres each were preempted in April, 1874. A host of other claims were located by others, but, from his knowledge of the ground, Searles and his crowd got on the best of it. Then some of those interested got word from Washington that the ground must be taken up under the placer-mining claim law—20 acres to the claim. Searles heard of this, by good luck, while in Los Angeles, and, in the procession that started out to stake off the marsh, he was well away at the lead.

The entire marsh was soon covered over with claims, and a great camp of prospectors—that held meetings and agreed, or were ready to fight among themselves, on various occasions—lived about the northerly side, a few miles from which was a cañon with good water in it. One night two of the crowd slipped out to jump John's claims. Everybody carried a six-shooter then, but when these two saw the old bear hunter with an armful of stakes, on their trail, they "coyoted out of that," to use a picturesque phrase coined on the plains.

Some of the claim-holders starved out and gave it up. A Los Angeles company spent $10,000 trying to make borax out of sulphate of soda, and were ready to lynch their expert from Nevada because the borax wouldn't come; but

SEARLES' GARDEN.

they found, in time to save the life of the poor fellow, that they had no borax on their claims. One Arthur Robottom, an Englishman, who has written a lot of borax papers, that are unconsciously comical, for various scientific journals, went to the marsh, and got his fingers in the mud. He arranged to buy 1,280 acres there, and a small establishment, managed by Mr. T. Dodge of San Francisco, was started. Something like 100 tons of borax was made, but the enterprise died from natural causes, and now only the establishment of which the San Bernardino Borax Mining Company is owner, and John W. Searles superintendent, is found on the marsh. The old hunter's pluck served him as well in his fight for business success as it did in his fight for life under the jaws of the grizzly.

CHAPTER XIV.

THE FIRST AMERICAN BORAX.

JANUARY 8th is an anniversary day in the history of borax, though one never observed as such, and seldom, if ever, thought of. It was on that day, in the year 1856, that Dr. John A. Veatch, while examining some water from a spring in Tehama County, Cal., eight miles east of Red Bluff, noticed that it contained borax, and that was the first discovery of borax in the United States. He had boiled the water until it was sufficiently concentrated to allow minerals in it to crystallize out, and borax was found crystallized on the side of the vessel. Thereupon, as might be expected of a Californian, Dr. Veatch became an enthusiastic prospector for borax.

There was no use in trying to turn the borax in the spring he had examined to commercial account, for it, with others near, was utilized in a health resort, wherein certain classes of invalids were finding much benefit. So the doctor began making inquiry, and soon learned of both springs and lakes where he thought borax might be found. It is not a little interesting to note, that the men of whom he made inquiry in this matter were the pioneer bear hunters of the State, the men who, like Col. Joel Lewis of Sacramento, went hunting grizzlies for the fun of the thing. Mr. Charles Fairfax was another hunter of the time, and both of these men, while hunting in the Coast Range near Clear Lake, in Lake County, say eighty miles north of San Francisco, had seen curious things—a mountain of "a white pulverulent substance," a rivulet that was totally unfit to drink from, and an Irishman with a story of a borax lake.

Early in September of the same year (1856), Dr. Veatch and Col. Lewis made a trip to the locality, lured chiefly by the tale of the lake of borax and the hill of "white pulverulent substance." The Irishman lived back in the mountains, and he told the colonel that an Englishman had told him that a lake somewhere in that country was full of borax, and the Englishman certainly knew what he was talking about, because he had once worked in a borax factory in England.

Going up the Sacramento River in a steamer, the doctor and the grizzly-killing colonel reached the town of Colusa, and there hired horses and rode back into the Coast Range. They found the mountain, the Irishman, and the borax lake, but missed the Englishman. The mountain was a big bank of sulphur, instead of borax or borate of any kind. The borax lake was full of salt, which was deposited in very beautiful bluish-red crystals. The Englishman was a fraud. The only real thing was the Irishman. He was just plain Irish.

However, the journey was by no means in vain. Aside from a lot of springs, where borax could be traced in the water, the doctor made a most important find. A Mr. Hawkins, who lived in the neighborhood, told the doctor about an "alkali lake," which "presented a rather peculiar appearance." He offered to serve as a guide. "After traveling a short distance and clambering to the narrow edge of an almost precipitous mountain ridge," they peered down the farther side upon "200 acres of fragrant mud of untold depth"—fragrant as sulphuretted hydrogen is. It was a hot day, but, undaunted by the infernal odor, the doctor slid down to the lake, "waded out knee-deep in its soapy margin, and filled a bottle with the most diabolical watery compound this side of the Dead Sea." He also gathered some incrustations from about the side of the lake.

Because it was from this lake that borax was first produced in commercial quantities in America, Dr. Veatch's description of it is worth repeating here. It was found among the hills on a cape projecting into the east side of Clear Lake. A ridge running across this cape "is composed of huge masses of rock, resembling pumice-stone, which floats like cork in the water." The whole neighborhood bears the marks of recent volcanic action—in fact, the action has not yet entirely ceased. The lake itself in winter covers about 200 acres, the water being at most but three or four feet deep. In summer there is but a fifty-acre space covered, and that less than a foot deep. The mud is well described as soapy—soft soapy, at that—with a depth of four feet. As afterward analyzed, the water, at an average stage, was found to hold .039 per cent. of solid matter in solution, of which 61.8 per cent. was carbonate of soda, 20.4 per cent. salt, and 17.8 per cent. borax. But that was not all nor even the most valuable borax in the lake. Within six months after Dr. Veatch's visit, very large crystals of borax were found in nests in the soapy mud of the lake. These crystals were of a green color and usually of perfect form, some of them being of remarkable size—from five to seven inches in length.

On this discovery being made, Dr. W. O. Ayers, who had become associated with Dr. Veatch in the investigation, sunk a coffer-dam three feet into the mud, where crystals of borax seemed to be, when the mud was prospected, in moderate quantity only, and took out 163 pounds of the crystals. Dr. Veatch, in a place where no crystals could be found by probing, took out 101 pounds. Thereat he concluded he could find 638,880 pounds of borax to the acre.

Borax, in those days, was worth something like 50 cents a pound. It was sold as a drug only, instead of as an article of common household use, as it is now.

A well sunk in the mud yielded eight gallons of water per minute—4,204,800 per annum. An elaborate calculation showed that this water could be boiled down, the borax crystallized out, and the product laid down in San Francisco at 3 cents per pound, gross cost; 3 cents from 50 cents leaves 47 cents per pound margin for profit, and the mine practically inexhaustible.

This was in 1856–57. In 1864, a company, with Dr. Ayers as working superintendent, began to make borax there. During the years that had elapsed between the discovery of this lake and its development, the importations of borax into the United States varied from $143,218 to $217,944 per year. In 1864 these importations fell to $8,984. California's borax lake supplied the rest. For four years the company grew rich, and then work ceased at this borax lake, because an artesian well, sunk as an experiment, proved such a gusher that it could not be plugged, and the lake was flooded, until it could not be profitably worked. Then another lake of the kind, known as Hachinhama, in the same locality, was developed, and proved a good investment, although its mud contained no naturally formed borax crystals. The solids held in solution there contained 75.4 per cent. of carbonate of soda (sal-soda), 8.3 per cent. of salt, and 16.3 of borax. There was such a preponderance of sal-soda that it had to be crystallized out with the borax and afterwards, being of greater solubility, it was washed away. In 1872, however, borate of lime deposits were found in Nevada, and for two years that was imported by the car-load, and mixed with the concentrated waters of the lake. When the mixture was complete, the carbonic acid gas released its soda and took up with the lime, which in turn released its boracic acid to take up with the abandoned soda—a change of partners that greatly enhanced the yield of borax at Hachinhama.

Nevertheless, the Nevada men had the weather gauge of Hachinhama—they found borax lying in fields. A wild competition began, prices went down, until borax, the one-time high-priced drug, was worth less than sugar; and down went the whole business, and the fortunes of about everybody in it. There is a trace of borax in the water of both the old lakes, but the crystals in the ore are gone, and the amount to be obtained from the waters of either will not pay for the fuel to boil it down.

CHAPTER XV.

STRUCK IT RICH ON THE MARSH.

THE people engaged in the borax business seem to have been so busy, each attending to his own particular marsh or claim, in the early days, that no attention was given to making historical records in the matter. At least, that is the impression I received in talking with them, for facts about the doings of early prospectors in Nevada, who are not now in the business, are very hard to get. However, it appears that as early as 1860, Dr. Veatch, the California borax pioneer, had found traces of the salt in Mono Lake, near the Nevada line; learning which, one William Troup, of Virginia City, began to look out for borax. In 1864, locations for salt, useful in reducing silver ores, were made on the marsh near Columbus, Nev., and a borate of lime, now well-known in the trade as ulexite, or cotton balls, was found there, though it attracted little attention then. In 1869, however, a teamster in the desert region south of Wadsworth found a cotton ball, which eventually got into the hands of men engaged in working Lake Tehama, where the borate of lime was very much needed in the production of borax. A prospecting party searched vainly for the place where it had been picked up.

In 1871, however, Mr. Troup, the Virginia City man previously mentioned, found some cotton balls forty-five miles southeast of Ragtown, near Salt Wells, and carrying some of them to Ragtown, he borrowed a wash-boiler of a

Mrs. Kenyon, put in the borate of lime with some water and carbonate of soda, boiled the stuff, and when the mixture had cooled and crystallized out, he had produced the finest borax ever made in Nevada.

At about the same time it appears that Troup re-found cotton balls on the marsh at Columbus. Within a year small plants had been erected both at Columbus and the Salt Wells finds, while 1,700 pounds of the cotton balls had been shipped from Columbus to a firm in San Francisco, and there worked. None of these deposits, however, created any special stir, even among those in the borax trade. Borax was worth 30 cents a pound by the car-load, and there was no thought of any change of moment in the condition of affairs, till the discovery of the Teels' Marsh deposit of crude borax set all the prospectors on the Coast wild on the subject, and started a rush that flooded the market, until the business was well-nigh ruined. How this find was made can best be told in the words of the finder, Mr. F. M. Smith, who is now president of the Pacific Coast Borax Company, because his story well illustrates Nevada life at that time. In an interview, he said:

"In the fall of 1872, I found myself among the wood camps about ten miles from Columbus, Nev. I had been following mining camps from Montana to Idaho, from California to Nevada, since April, 1867. During this time, like most men who follow mining camps, I had engaged in nearly every vocation pertaining to mining. The chief object of the camp follower or prospector is always to find a good mine. In this enticing, but most delusive pursuit, all available employments are accepted for the time, but only as incidental to the one object. I had recently been engaged in teaming, contracting for the delivery of wood to the mills, and timber to the mines.

"I then owned two or three wood-ranches, and a band

of pack-animals, and I had also accumulated the usual number of wild-cat claims to mines.

"Just before the discovery of Teels' Marsh, I had bought a wood-ranch of a fellow woodsman, and, by the help of a wood-chopper in my employ, had erected a good, comfortable cabin in a narrow gulch, commanding a fine view of the outlying country. The view from the adjacent timber

RESIDENCE OF F. M. SMITH, TEELS, MARCH, 1873.

included the Columbus Borax Marsh, which was being worked, also Teels' Marsh, of which nothing was then known as a borax deposit.

"I should say that an alkali marsh, so termed, differs from an Eastern marsh. It is not watery, and not necessarily soft. It is a dry lagoon, or surface incrustation of alkali in some of its forms—soda, salt, or borax. It has a light, whitish appearance, often covering large areas, but quite variable in surface color. My prospecting had included borax claims, but I had very little knowledge of the business.

"One morning, immediately after my cabin was completed, I heard some one chopping wood near by. Upon going out to examine, I found a Mexican chopping in the timber designated and described as my property. He refused to leave when I ordered him off the premises, and I laid hands upon the ax. He was reinforced by another Mexican, and a white employe. I soon found that they intended to forcibly dispossess me. The nearest officer or Court of Justice was at Aurora, the county-seat, fifty miles distant, and the only convenient or satisfactory appeal was to the rifle. I had no weapons, and knew short arms were of little use. I went immediately to a fellow woodsman, two miles distant, who owned a Spencer rifle, and found that the weapon was at Columbus, ten miles away. He gave me an order for it, and I returned that night to my cabin with the carbine, and only four cartridges, all I could get in town to fit it. I found the Mexicans had reinforced their gang, and had been chopping down a fine cluster of pine trees in front of my cabin door, and had several cords of wood neatly piled up in place of the trees I had so admired. My chopper told me they were in full force, and intended to bring in a pack-train in the morning, to carry out the wood they had chopped.

"We breakfasted early the next morning, and my chopper took his ax, and went to the timber some distance away, to chop. I knew enough of him not to count upon him for any assistance in maintaining my rights, and did not ask him to help me.

"I soon heard the tinkling of the bell worn by the bell-mare which always leads the train, and took my position on the hill-side, just out of close shooting range with small arms, and some 125 yards from the wood-pile. Then the train appeared, and I counted twenty-one pack-animals, four Mexicans, one Indian, and one white man. Upon reaching the timber, two Mexicans alighted, and took one of the animals and led it to the pile of wood, and began to load.

"In loading the aparajo, it is customary to put on a big stick first. This wood was green and heavy, and it took two men to handle the larger sticks.

"They had seen me as they passed up the hill, but I had said nothing to them until they began to load, when I called out, 'Hold on there, don't load that wood,' and took aim with my rifle. At this, one of the Mexicans dropped his end of the log, and showed no inclination to take it up again, though urged by his companion to do so. Both replied to my challenge with abundant oaths. I persisted in my demand for them to 'clear out,' and, after plenty of cursing, one of the party began to approach, saying he wanted to talk with me.

"As he stepped forward, I took deliberate aim, and again bade him 'hold on.' He stopped advancing, but did not stop swearing. This was repeated several times. I knew I had no ammunition to waste, and that at close quarters with their knives and pistols I stood no chance, even were their numbers less.

"They finally offered to compromise by taking the wood they had chopped. I replied, they had no business to chop it on my claim, and no right to it, now it was chopped. I stood my ground with the rifle at my shoulder, and, as a final result, the train of twenty-one mules, four Mexicans, and two helpers, turned about and retired empty-handed, leaving me in possession.

"The next day they sent a delegation from Columbus to

propose a compromise, but, as I had no compromise to make, they also retired, and I had no further difficulty from that source, though, it is true, for some time an assault on my cabin was not unexpected.

"Not many days later I did have a night alarm. It was the visitation of the earthquake which proved so disastrous to the Lone Pine District, where some thirty lives were lost. The shock was very severe in the hills. I was alone

THE FIRST BOILING PAN IN USE AT TEELS' MARSH.

in my cabin that night, and woke at the violent shaking, and the general rattling down of all loose articles. I at once concluded the Mexicans had made an attack, and sprang to my feet, pistol in hand, and shouted, as I had done when they attempted to load the wood, 'Hold on, there!'

"The shaking did not hold up at my command, however, though it grew perceptibly less, and I soon fully understood the nature of the attack. The tremblings continued at intervals for several days.

"It was a dry season, and the alkali areas were more than ordinarily extensive. From the hill-tops I could see the gleaming-white Teels' Marsh, and taking two choppers one day, I visited the marsh, and found a heavy incrustation, which, on testing, seemed rich in borax. It afterward appeared that we had chanced to step upon the richest portion of the marsh first.

"We made a preliminary location that day, and gathered samples, which, on the day after, I carried to an assayer at Columbus. We were so impressed with the appearance of the marsh, that we took a small supply of provisions and pack-animals from the wood-camp, and returned to the ground, made a dry-camp, and at once located several thousand acres, most of which afterward proved to be worthless.

"After starting the men at work, I went to visit the wood-ranches, and then on to Columbus. On my way thither, the certificate of analysis, which pronounced it the finest specimen of borate of soda that had been found at that time, was handed to me. In Columbus I enlisted two associates, with one of whom I laid in fresh supplies, and started back for Teels' Marsh. Long after dark, at a point some nine miles from the marsh-camp, we found the Mexican who had been slowest to drop the log when the attempt was made to run off my wood. He had been out hunting stock all day, and was glad to get the benefit of the camp we were ready to make, for he had no provisions, while we had plenty. He was on friendly terms, now, and made from our flour tortillas for our supper, the first I had ever eaten.

"I felt so anxious to anticipate any attempt to locate the deposit which the favorable assay might inspire, that, guided by the camp-fire of the helper I had left on the marsh, I pressed on to his camp, reaching there after midnight. It was a good thing I did so, for I found a friend of the assayer on the ground. He had been sent out to locate borax deposits. He had found my men, and remained with them till I came in. Pretty soon he asked me where Teels' Marsh was, and I told him I knew an alkali flat about twenty miles from there, and, in the morning, I would send a man with him who was familiar with the country. Next morning he left very early, accompanied by one of my choppers. They were gone three days. It is scarcely necessary to say that, by the time of their return, I had the property well located.

"Some days later, the assayer himself put in an appearance, and I staked him out a good location, and put him on it, which was very pleasing to him, though perhaps scarcely deserved.

"Borax lands had heretofore been located under the saline laws of the State, one locator taking 160 acres. Soon after the date of my discovery—the fall of 1872—Commissioner Drummond had decided that borax land should be located as placer claims, allowing only twenty acres to the individual. This made the work of locating more difficult and expensive. Besides, borax land is usually very spotted, varying greatly in richness—thin, mixed, and only a small portion profitable to work. This feature rendered it necessary to thoroughly prospect, to make a careful selection, and to cover as much land that was valuable as possible, to warrant erecting a plant for the manufacture. It is true, borax was then worth 30 cents a pound by the car-load, but it soon dropped to one-third that, while grain cost $140 a ton, and hay $60 at Columbus, twenty-five miles distant, with no roads for transportation to my camp. Wood and labor were both expensive, and the product was far from market, with no suitable eans of conveyance.

"We were compelled to re-locate our property, and to secure the co-operation of all our friends, using their names as locators, and soon thereafter get deeds from them for their claims. The wood-choppers associated with me had been following mining camps too long, and had 'lost their grip.' Though fully aware of the richness of the deposit, they did not put a high value upon their interests, and perhaps it was true that the less money they had the better off they were. I made a trade with one of them, giving him a wood-ranch for his interest. The other wanted $300 for his claim, which I bought and turned over to one I supposed would be a desirable party. He passed a portion of the interest to other and unfriendly hands. This was only re-secured after long delay and much trouble. At one time as many as seven locators came onto the marsh and took forcible possession, standing with guns in hand behind little monuments they had built on the alkali flats. Their numbers enabled them to relieve each other, and I was out-numbered. I had to appeal to the courts for their ejectment.

"As soon as possible, I made arrangements, through my brother, with a Chicago company, to put up a plant, and the production of borax was begun. Borax was then little known beyond the blacksmiths and druggists, and was costly, druggists selling it at twenty-five cents per ounce. The total consumption in the United States, we soon learned, reached only about 600 tons per annum. Before we could get any into market, it fell to 10 cents a pound, which was all we realized.

"I have lived to see borax become an important article of commerce, and, by reason of the great reduction in price, a household staple of universal use. It is now cheaply sold in convenient packages by all grocers. Teels' Marsh has been operated almost continuously ever since. This property, for a long period of years, practically controlled the borax market, and was the richest and most extensive of all discoveries. The total product of the marsh has probably reached 17,000 tons.

"I have been identified with the borax business from that date to the present time, twenty years. Smith Brothers finally obtained sole control of Teels' Marsh, buying out over one hundred locators, and clearing up all adverse claims. It subsequently passed into my individual ownership, and was transferred to the Pacific Coast Borax Company about two years ago."

PACIFIC COAST BORAX CO.'S WORKS, ALAMEDA, CAL.

CHAPTER XVI.

BORAX IN THE MARSHES AND MINES.

I AM bound to say that, when starting on my journey through the desert regions included in these sketches, I was animated chiefly by a desire to see the phases of life to be found there. Nevertheless, I was curious to see borax, and borax materials, as they were found in the earth, and if what has so far been told of the country and its people has turned the curiosity of the reader to the subject as mine was turned, a chapter of descriptions of borax deposits will not be inappropriate at this point.

As told elsewhere, Teels' Marsh was the first deposit made commercially profitable in the desert region. Any good map of Nevada will show the location of Teels' Marsh in Esmeralda County, Nev., and the Rhodes, the Columbus, and the Fish Lake deposits are found in the same county and not many miles away. All are marked on the ordinary maps. The marshes in California include one in Death Valley, the one belonging to the San Bernardino Borax Mining Company (which on some maps of San Bernardino County is marked "Searles Brothers Marsh"), and one in Saline Valley, Inyo County, which I did not see. There is also a borate mine in Oregon, which I did not visit.

In general appearance—that is, to one not familiar with the details—these borax marshes are exactly alike. There is a bowl of a valley surrounded by picturesque, if barren, mountains, and at the bottom of this bowl is a vast deposit that looks like wáter, or salt, or dirty snow, or chalk, according to the conditions of the air and the distance of the spectator. When one walks across the marsh, he finds it covered for the most part with a sandy-looking crust through which the feet generally break. Below this crust there is commonly clay, more or less wet, and in places water and slime of unfordable depth. A Piute, who tried to ride across Teels' Marsh once, lost his pony in the mire, and, in working all these marshes, many animals have narrowly escaped similar fates. The marshes are, beyond doubt, the bottoms of lakes, now dried up—they are, in some cases, real lakes now, during wet seasons, though the water is rarely more than a foot or so deep. In the case of the San Bernardino Company's marsh there is a very distinct water-mark on the mountains on nearly all sides, at a height of about 600 feet above the present surface, while other and fainter water-lines are seen higher up. No other marsh shows such distinct water lines as this one, but traces of the old lakes can be found at all of them. There is even a similarity in the form of the marshes. They are nearly circular, save in Death Valley and the Amargosa Valley. The peculiarity of those two deposits will be dwelt upon further on in this chapter. Because of this oval or circular form, and because of the volcanic character of the entire region round about each, it is supposed that each lake was formed in what had previously been the crater of a volcano. Moreover, they were all the sinks of mountain streams—there was no communication between them and the sea, but, because of changes in the climate and in the amount of the rain-fall, the water supply grew less and less, and eventually each lake became a marsh. There are a few lakes in the region, such as Mono, Owens, and Walker's, that have not dried up, but these all had streams rising in mountains that were snow-covered every year—in fact, the feeding streams of each of the three mentioned originated in the Sierras.

Beyond the general characteristics mentioned, there is much difference between the marshes, and particularly in the deposits of borax materials. These differences are so great, in fact, that no one can say just where the borates came from, how they were formed, or how deposited. This much is known: Borax, in the language of the chemist, is the bi-borate of soda--it is a compound of boracic acid and a small proportion of soda. Boracic acid is formed naturally in some volcanic regions, perhaps as certain other products are formed—the sulphurous gases, for instance. This boracic acid, in gaseous form or in solution, came rising from subterranean laboratories through certain other substances, and formed compounds with them. It came in contact with lime, and formed borate of lime, and with soda, and formed the borate of soda. These combinations were formed under different circumstances, and therefore the compounds differ from each other in appearance, but just what these circumstances were, nobody knows.

My first view of a borate was at Rhodes' Marsh. It was in a form known as cotton balls. The scientific name is ulexite, and it is a borate of lime. I saw cotton balls as small as a pin-head at Teels' Marsh, and as large as a good old-fashioned pumpkin in Death Valley. They were found imbedded in a tough clay, as a rule, along with sulphate of soda, magnesia, common salt, etc., but down on the grounds of the Columbus Borax Company, at Columbus, they were in a sandy soil, which the workmen were sifting through a screen, in order to eliminate the cotton balls. When first taken from the earth they are easily broken apart in the hand, showing a silky-white fiber, very beautiful to the eye. After the exposure to the air, they become so hard that they are sometimes ground in a mill before being made into a solution. In the ground of the Pacific Coast Borax Company, on the Columbus Marsh, this cotton ball material is found decomposed and mixed with a sandy loam, which it whitens, so that the workmen readily shovel away the surrounding earth, leaving the beds of valuable material exposed.

Now, although these cotton balls are unquestionably formed in the mud of an old-time lake, it is not possible to say how they were formed. They occur in beds, and nests, and singly as well. They are in no case distributed all around the circumference of the old lakes, nor are they all on one level. Moreover, they are found in the Esmeralda County (Nev.) marshes at an average elevation of more than 5,000 feet above the sea, and in Death Valley at the sea level and below it. More curious still, they are found in the Furnace Creek cañon of the Funeral Mountains, 1,500 feet in altitude above, and twenty or twenty-five miles away in a straight line from the cotton balls in the Death Valley Marsh, with huge mountains between. On the other hand, there are no cotton balls in the San Bernardino Company's Marsh west of the Slate Range, nor is there any borate of lime there in any form.

My next view of a borate deposit, after visiting Rhodes' Marsh, was at Teels' Marsh. There the chief deposit was the borate of soda—crude borax, combined with alkalis. It once lay in wide beds, several inches thick, white, and, to the uneducated eye, like soda. It had sand in it, deposited there by sand-storms, but it needed only to be boiled up in water and crystallized out, to be ready for market as the borax of commerce. These beds were long ago worked off, and then appeared a new feature of the borate of soda deposits. The borax, as the workmen said, grew again, though not to a satisfactory extent.

Wherever there is a deposit of the borate of soda, the

crust, formed on the deposit, continually increases in quantity—it oozes up from below and forms a crust over the surface. The carbonate of soda, found on all these marshes, does the same thing, and so does the salt, which is also found on the marshes. In fact, these

VIEW OF MONTE BLANCO.

two grow much more rapidly than the borate does. The carbonate of soda looks very much like the borate to the uneducated eye—they are both whitish, sandy crusts, that look somewhat like tiny waves on a rumpled, muddy lake, but there is a difference between the two crusts that is readily detected after some experience. At the San Bernardino Company's marsh, this deposit forms in a much harder crust than can be found on any other marsh. The workmen have to use picks in breaking it up. In Death Valley, along the marshes, are wide fields of this crust-like borate of soda, but, what is more curious, is the fact that certain clay buttes on the east side of the valley, pretty well up toward the north end, are also covered with the same kind of a deposit. Over in the Amargosa Valley the deposits are found in localities—like those in Death Valley—down on the low grounds of the flat and up on top of clay buttes.

But the most remarkable deposit of crude borate of soda is at what is called Monte Blanco, on the south side of Furnace Creek Cañon, in the Funeral Mountains. Not only is this deposit far removed from any marsh—it is a hill of crude borate of soda, surrounded by other hills. There a wedge-shaped peak, perhaps a thousand feet high, the top and sides of which are covered with borate of soda, in the form of a white sandy powder several feet thick. There is no guessing at the amount, nor is an estimate of the amount of any value; but how did it get there among the mountains?

There is still another form of crude borax called tincal, of which mention is made in the chapter on the first discovery of California borax. This tincal is a crystallized borax—it can be used as it comes from the ground in certain arts; it could be mixed with lime, for instance, in making fire-proof cement or mortar, or as flux in melting or soldering metals. The first borax shipped from the San Bernardino Company's Marsh was in this form. Tincal is also found at the Rhodes Marsh. The deposits of crystals are in scattered nests and beds. At the San Bernardino marsh, a space of 300 acres was found to be full of the deposits, and they were to a great extent dug out. Then it was discovered, after a time, that tincal crystals would grow. The men who dug out the crystals had to work in an offensive mixture of mud and water. In digging them out, holes were left in the mud, and these holes filled with water. There was, apparently, a constant ingress of boracic acid in solution into these holes, and this, coming in contact with soda in solution, formed crystals of borax, which were deposited in the mud. Supt. Searles found borax crystallized on a bit of brush in a well-hole. So a search for new-formed crystals was instituted, and at the time of my visit a pile, containing perhaps 1,500 tons, had been dug out. These borax crystals were mixed with crystals of salt and carbonate of soda, but both of these substances, being more easily dissolved than borax, can be washed out, and then the borax will be refined by solution and crystallization.

Acting on the hint found in the old holes, the superintendent had a great vat, perhaps a half-acre in extent, constructed, and into this the natural borax solution is run to form its crystals. This is a unique feature in the production of borax.

For a long time after the discovery of the borax materials in Nevada and in San Bernardino County, California, the people in the business believed that the borate of soda crusts on the marshes, the tincal, and the cotton balls were the only commercially available deposits in nature, and it happened that even after the work on the borax marsh in Death Valley was begun, the searchers for other deposits drove, when prospecting, over an extensive ledge of material that really assayed higher than the ones they were working. It lay in the road out of Death Valley, was crushed by the wheels of buckboards, and was knocked out of the path by the prospector, because it obstructed his way. But this condition of affairs did not last long. It happened that very valuable deposits of silver ore had

BORAX AS A "QUARTZ-MINING PROPOSITION."

been found in the Calico Mountains, and, in consequence, the whole range was pretty soon covered with wild-cat claims. Among the many bright-colored strata and ledges to be found in these mountains was one that was snow-white, and composed of an unknown material. On casual examination, it was found to contain radiating crystals of singular beauty, but not wholly unlike some other crystals to be found in volcanic regions. Eventually, because this rock was different from any other, some samples were analyzed, and behold! Here was borax in a form never dreamed of. The curious formation was a borate of lime, which showed in the analysis more boracic acid than cotton balls. It was called colemanite, after Mr. W. T. Coleman, who was associated with Mr. F. M. Smith in the discovery.

Hitherto the gathering of borax materials had been, to adopt the California term, a placer-mining proposition. The workmen, in gathering it, shoveled it up into windrows on the marsh, and loaded it into wagons. that it might be hauled to the refining works. Now they were to take hold of a genuine quartz-mining proposition. The new material was a well-defined layer or ledge, and it could be mined only as any ore might be. It cropped out along the peaks and ridges in such a way that it could be readily traced, and the stretch eventually patented under the mining law was $1\frac{5}{8}$ miles long. The ledge was found to average perhaps six feet in thickness, but to attempt to describe its pitches, angles, dips, etc., would prove futile as well as uninteresting. In trying to get an idea of the deposit's characteristics, I climbed over the broken hillocks for half a day, and then gave it up.

The deposit, although of very great extent, is not without its drawbacks, for it is located in the midst of the roughest kind of a desert mountain range, and the road that was constructed to it has more crooks and pitches than the streak of chain lightning. Moreover, all the water for use about the mine has to be purchased of the Atlantic & Pacific Railroad Company at Daggett, and hauled to the mine. Then another drawback is the utter lack of fuel, and, in consequence of this fact, the Pacific Coast Borax Company, who own the mine, have to ship the crude material as it comes from the mine to San Francisco, to be manufactured into borax.

It was a deposit of borate that obstructed the pathway of the searchers for borax in their journey to the eastward out of Death Valley. Another and much larger deposit than the one in the Calico Range is found right beside the mountain of crude borate of soda at what is called Monte Blanco, in a branch of the Furnace Creek cañon.

Beside both the Calico and Funeral mountain deposits of colemanite, another deposit, known as pandermite, is found. This is borate of lime, rich in boracic acid. It looks some like marble of fine grain, and some like unslaked lime. It is found in large quantities, but is not worked at present. In fact, none of the Death Valley regions now are worked. The long distance over which the supplies and the product had to be hauled, the lack of fuel, and the terrific heat of summer combined to make the deposits there unprofitable. The same may be said of the Amargosa Valley deposit. These vast stores of borate are now waiting for the developing influence of some eastern railroad seeking a share of the Pacific Coast trade. Once a railroad brings cheap fuel to Death Valley, business will boom there—though the statement that the chief want of the hottest spot on earth is cheap fuel, strikes the average reader as a trifle weird.

In Prof. Hank's report on borax deposits, issued by the California State Mining Bureau, no less than twenty-two natural borates are mentioned, of which one, a borate of lime called hayesene, is said to be found at Bergen Hill,

THE NARROWS—"A ROAD WITH CROOKS AND PITCHES."

N. J. Another borate that, because of the peculiarity of its crystals and its rarity, is much prized by collectors of minerals, is called hanksite, after Prof. Hanks. It is found only at the San Bernardino Company's marsh, as is also still another form, identified as such during the year

IN THE BORAX MINE—BY FLASH-LIGHT.

1891, and named searlesite, after the superintendent of the works who analyzed it.

Although, as said, no one knows just how the borates were formed, the borax sharps make a guess that all of them were made in water and mud, including the colemanite, and the borate of soda, and the pandermite found in the Calico and the Funeral mountains. This guess involves wonderful, but by no means incredible, geological changes in the surface of the earth, for it supposes that these deposits were once at the bottom of lakes, that the rains of many years washed down various kinds of sediment on top of them, that volcanoes spewed molten lava over them, and that then came a convulsion of nature that ripped open the surface of the earth, split apart the bowls of the lakes and tossed up vast surfaces in breaking waves with the spray of sand in clouds, and when all had settled down again, mountains had taken the place of bowls and valleys —the great lakes that once included Death Valley, the region of the Funeral Mountains, and the lower end of the Amargosa Valley, had been divided by the upheaval in its midst of the many-colored Funeral Range. They guess that colemanite may have been cotton balls once, and that cotton balls were changed by pressure, or heat, or both, or something else into the new form. They guess that if borates are found on one side of a marsh and not on the other, it is because the boracic acid formed in a volcanic laboratory found a vent to leak up through in that vicinity and nowhere else. They guess the pandermite, found in thin streaks running in every direction through certain rocks in the Funeral Mountains, was worked into them when the rocks and the pandermite were in a plastic state. On the whole, there is probably no better place on earth for a teacher of geology, who wishes to illustrate to a class of students the known and the guessed-at facts in the formation of the earth, than the deposits of borax and the mountains round about in the deserts of Nevada and California.

The following article, clipped from a San Francisco newspaper, is inserted here, because it is a fair sample of a very large number of articles that have been printed since the exciting days of 1872. It appeared while I was making the tour of the deserts:

"Large borax beds, which are now attracting attention, are situated sixty-five miles due east of Ellensburg, in Douglas County, and are at present twenty-one miles from the nearest railroad point, says the *Capital.* The deposit has the appearance of a lake, a mile and a half long by half a mile wide, except that instead of containing water it is a solid deposit of borax, which is eight and one-half feet thick, resting on clay, which, in turn, overlies a bed-rock of

IN THE BORATE MINE—BY FLASH-LIGHT.

slate. The borax is solid, and almost as transparent as ice, and can be easily cut with an ax, or it can be sawed out in blocks of any size that might be desired, and it is so easily cut that one man could take out a vast amount in a day. On exposure to the air it slacks, in which state it looks like lime. In this process, however, none of its strength is lost, but if put in water it seemingly dissolves, but only to resolve itself into its original icy consistency again. While the dimensions of this lake and the depth of the deposit would seem to give a fair idea of the quantity contained there, such is not the case, and, instead of a computation of the supply being possible, it can truly be said that the supply is utterly inexhaustible, for the reason that the substance bubbles up, milk-warm, from the bottom, so fast that were a hole, say, for instance, ten feet square, cut out, it would be refilled in forty-eight hours."

The fact is, this so-called borax was epsom salts, commercially valueless. The number of epsom salts and salsoda fields that have been mistaken for borax is almost legion, while not every real borax deposit is commercially valuable. Competition in the manufacture long since brought the price of the product down to a point where any borax proposition, to use the California term, must be one of exceptionally rich material and well located, if it is to pay. The location is, indeed, the main thing, for the deposits in Death Valley are the richest in the world, but, because so far from market and fuel, are allowed to lie unused.

BREAKING CAMP.

AT THE MINE, READY TO LOAD.

CHAPTER XVII.

A CHEMICAL ROMANCE.

MANUFACTURING borax from the crude materials gathered in the deserts is one of the romances of chemistry; the description of the process is a story of fickle affinities, wherein, as commonly happens with such affinities elsewhere, the two fickle ones unite in a waste combination, while the deserted ones get together to make a respectable and valuable product.

The crude material from which most of the borax is manufactured is the borate of lime, a combination of boracic acid and lime. As told elsewhere, it is mined in the Calico Mountains, in San Bernardino County, Cal., as a crystal called colemanite, and from the Rhodes, Teels, and Columbus marshes, in Esmeralda County, Nev., gathered as a silky substance, somewhat like asphalt, called cotton balls.

The borate is hauled from the mine to Daggett, on the Atlantic & Pacific Railroad, and shipped by the car load to a chemical factory at Alameda. Some idea of the extent of this factory may be had from the dimensions: Thus, one part is of concrete, 40 x 230 feet; another is of frame, 80 x 170; a third is 26 x 110, and all three are three stories high. There is a one-story building, 80 x 145, and a shed, 30 x 110.

The crude stuff, in lumps of all sizes fit to handle, is

THE CRUSHER, MILLS, AND ELEVATORS.

dumped off the cars into the mill. It is there fed into the iron jaws of a breaker, which reduces it to lumps smaller than a walnut. Thence it goes to a coffee-mill sort of a machine that reduces it to a sand. Thence it goes to a stack of rolls and on to burrs, in a process precisely like that used in making flour out of wheat, and ends up by going through bolting-reels covered with silk, so that a powder as fine and smooth as flour is obtained.

Meantime, a lot of carbonate of soda (sal-soda) has also been powdered. This sal-soda is a natural product of Owens Lake, Inyo County, California, which is obtained by solar evaporation of the water.

Having the two powders ready, the borate is dumped into a hopper by a common elevator and conveyor, and the soda is dumped in with it. Under the hopper is an iron boiler. This boiler has a great paddle-wheel inside, revolving near the bottom, and it is pretty nearly filled with water, or a liquor to be mentioned further on. The water is set boiling by a steam-coil, the two powders are dumped in and the paddle-wheel set going to stir them up.

Thereupon, the heat and the turmoil upset the existing affinities completely. The carbonic acid in the carbonate of soda drops the soda very quickly, and jumps at the lime in the borate of lime. The lime yields, and the boracic acid, being released, straightway takes up with the abandoned soda.

In the new unions, the carbonic acid and lime unite to form carbonate of lime, an insoluble substance that very readily precipitates when the paddle-wheel stops agitating it. The boracic acid and soda form bi-borate of soda, and that is the chemists' name for borax. The borax remains in solution, and the solution is run into tanks where the borax can crystallize out. The sediment remaining in the boiler is washed and re-washed, and eventually put into a

THE CRYSTALLIZING VATS.

press and squeezed, in order to get the last drop of borax solution from it.

In the tanks, where the solution is allowed to cool and settle, the borax forms in dark-colored crystals, called precipitates, that is pretty good borax, but not quite suitable for commerce. When it has all crystallized out, the liquor is drawn off and pumped back into the boilers, to be used over and over in upsetting affinities, until at last it gets as foul as a go-between might expect to become, and is thrown aside.

The impure borax is made into a new solution in tanks of hot water, and the solution is run into vats to cool. These tanks may be of any shape, but are generally round tanks, between five and six feet deep, and of about the same diameter. In these tanks stout wires are suspended, and the borax crystallizes on the wires, and the sides and bottoms of the tanks. When the wires are taken out, crystal-covered, they look like sticks of rock-candy, five feet long and five inches thick. This borax is the borax found on sale in the groceries, save only that it is commonly sold as a powder—the crystals have to be ground in a mill. For use in the arts, it is sold in crystals.

The borax on the bottoms of the crystallizing tanks sometimes has a stain in it, but it is sold for a certain purpose—for use as a flux, for instance—in which a stain is of no consequence. Of course it brings a less price than the absolutely pure.

The manufacture of borax from the cotton-ball borate of lime is chemically the same as when colemanite is used, but there is a difference in the plants and practical workings. At the Columbus, the Teels, and Rhodes marshes, the cotton balls are gathered and mixed with carbonate of soda, but the boiling is done in open vats, where the material can be stirred by hand. Moreover, there is a large

THE CRYSTAL BORAX.

amount of sand and clay mixed with the cotton balls, and this, when the transformation is complete, is washed out to a tailings-heap, which can eventually be worked over for the borax that remains in the sediment.

The works of the Rhodes Marsh Company and of the Columbus Borax Company heat their vats by steam boilers, under which wood is burned. There are, in addition, small outfits around the Columbus, the Teels, and the Fish Lakes marshes, where the vats are heated by sage brush fires, applied directly. It is a question of economy in fuel, and that is a very necessary question to consider, where soft wood costs from $10 to $12 a cord.

The marsh of the San Bernardino Borax Mining Company produces crude borax, and no borate of lime. The borax forms in a crust of sand and clay, over the surface of a part of the marsh. It is shoveled into windrows, hauled to the works, where a solution is made of it in tanks, heated by steam from boilers heated by a crude petroleum flame. When boiled up, the solution is allowed to settle, and is then drawn off into large cement vats, where it cools, and the borax crystallizes out in a form not quite pure enough for the ordinary market, and it is therefore re-dissolved and re-crystallized, when a beautiful product is had. The residuum is run from the dissolving pans through a series of vats devised to save such borax as remains in it.

This marsh also produces great quantities of tincal, or the natural crystal of borax, but the tincal is dissolved and re-crystallized before it is sent to market. These works are more than seventy miles from a railroad, and are located on an arid desert, but are models of mechanical construction, which, although of small interest to the ordinary reader, would be worth a visit from any one versed in chemical manufactures.

CHAPTER XVIII.

SKETCHES OF BORAX IN OTHER COUNTRIES.

STORIES of ancient Rome, when such sporting characters as Nero found delight in deadly encounters between gladiators, afford the first authentic reference to borax. Sir Edward Bulwer-Lytton says in "The Last Days of Pompeii," "Borax was largely used by Nero and his slaves," and "Panza deeply regretted that he was not rich enough to buy borax to cover the arena after the death of the combatants, after the fight between Lyden and Tetraides."

Some of the borax sharps have thought that borax was used by Nero to deodorize gladiatorial battle-grounds. I do not think so. Nero was a man who loved blood. The odor of it was sweet perfume to his nostrils; the sight of it was a delight to his eye. He used borax, not as a disinfectant, but to preserve the relics of the fight in all their ghastly reality, that he might come to gaze upon the scene and live over again the excitement of the contest.

Because there is no other reference worth mentioning in early or ancient history to borax, the statement of Prof. Henry G. Hanks, that "the early history of borax is vague and uncertain," seems to be justified. It was not, in fact, until about the time of the American Revolution, that writers began to pay much attention to borax. What they didn't know but published about borax at that time, forms a good parallel to what some modern writers didn't know but published about Death Valley. In his "Elements of Natural History and Chemistry," published in London in 1790, M. Fourcroy says: "We get this salt from the East Indies * * * * M. La Pierre, apothecary in Paris, has imagined it to be formed in a mixture of soap-suds with dirty kitchen water, which a certain individual preserves in a kind of ditch, obtaining from it, at the end of a certain time, genuine borax. But this fact, though first communicated to the public ten years ago, has not yet received confirmation."

However, it appears that some traveler afterward crossed the Himalaya Mountains into Thibet, and found there a series of lakes that somewhat resemble the lakes of California and Nevada in which borax has been found. They were in a basin about one thousand miles long, stretching away from Leh and Ladak, east by south to Lassa, all of which places may be found on an ordinary good map of India. The lakes were without outlet, and, therefore, full of salts.

Digging in the mud of these lakes, the natives found "greenish masses of opaque crystals, of a greasy feeling," and these crystals, having found their way into the hands of Europeans, were readily salable. They were called tincal by Europeans, and to this day the crystals of borax, formed naturally, are called tincal. Natives of that country called the stuff baurach, and that is said to be the origin of the word borax. When tincal was first found in the lakes of Thibet, and when or how the stuff was found to have a value, were questions which the borax experts, with whom I talked, could not answer. May be Nero got his borax for preserving the features of the bloody battle-grounds of his gladiators from Thibet.

The tincal, having been gathered at the lakes, was carried to Lassa, and there "bartered for cowrie shells, Sheffield cutlery, and Birmingham ware. It was then sold to the Kassawaris and Khampos traders," who carried it over the Himalayas, so that it eventually reached Calcutta.

The method of transporting this crude borax over the Himalaya passes was in marked contrast with the American method of transporting borax from Death Valley over the Panamints to market. The Yankee freighters hauled 40,000 pounds at a load. The Kassawaris trader bought a flock of sheep, lashed a package of from twenty to forty pounds of borax on the back of each animal, and away he went. Some animals carried as high as fifty pounds. Food for the sheep was found by the wayside, though in places trees had to be cut, that the sheep might browse on their leaves. When a sheep died, the drivers ate its flesh, and spun the wool into yarn, each man carrying a distaff for the purpose. The load of the dead animals was distributed among the living.

Lassa was 500 miles from Calcutta, but in recent years a railroad has cut off 300 miles of this distance. From Calcutta, the crude borax went to Liverpool, where it was refined and used in glazing pottery.

This Thibet tincal was the only form of borax known in England until the year 1742, when an Italian source of supply was discovered. In that year, one Targioni Tozzetti, a traveler of scientific attainments, got down into a section of Tuscany, Italy, about sixty-five miles southeast of the city of Leghorn. "He relates," says W. P. Jervis, of the Museum of Turin, "how he took a stroll through the valley which stretches southeast from Monte Cerboli, and reached the little torrent Possera. All around him was a scene of desolation well fitted to strike dismay on the ignorant, but eminently suited to the contemplative mind of the naturalist, to whom the most dreary plains and barren rocks yield ample subject for useful and agreeable study."

Here he began to examine the gulfs and chasms about him. Rumbling noises could be heard in their depths, and disagreeable odors filled the air. His guide told him that flames could often be seen in some of the cavities at night. Near by the chasm he found little muddy pools of blue water, "boiling vehemently," the escaping vapor producing swelling bubbles, which, on bursting, let clouds of white vapor, "smelling strongly of rotten eggs," rise in the air. The ground was soft, and crumbled under his feet.

The whole of the valley was full of these little lagoons, but, because of cracks and cross-fissures, no estimate of the number could be made. Occasionally these puddles were filled to overflowing by rain, and then the hot water ran into the Possesa and killed the fish for a considerable distance down stream, while during cloudy weather, and what we would call muggy weather, the rumblings redoubled in their fury. Near Castelnuova he found the pools increasing in number, though some old ones dried up and emitted steam and gases only at intervals.

The forming of new pools was alarming and dangerous to the people of the vicinity. A farm house near Castelnuova, built 200 years before, was one day undermined. A pool suddenly formed under the kitchen. The inhabitants fled from the ancestral hall, and it soon went to pieces. A story was told of a hog drover who suddenly lost thirty-nine of a bunch of forty pigs. Sheep occasionally tumbled in, and were quickly boiled to pieces. One man lost his life. He was working in an alabaster pit, when a vent was formed in the bottom, admitting hot carbonic acid gas from below. He screamed for help and his comrades strove to help him out, but before he reached the top the gas overcome him, and, relaxing his grip on the rope, he tumbled back into the pit and died.

But the region was not without its amenities. The peasants, whose chief article of diet in place of bread was the chestnut, managed to roast their crops by the heat of the dry pools, while the animals of the country round about sought the place for its heat in winter, and to escape flies and mosquitoes in summer.

Tozzetti's account of what he saw in 1742 led a chemist named Hoeffer, in the employ of the Duke of Tuscany, to examine the region in 1777, and he discovered that boracic acid (which, combined with soda, makes borax) was held in solution in the "muddy blue water" at Monte Potondo and Castelnuova. In 1779, "Prof. Mascagni, well-known for his researches on the lymphatic system," confirmed Hoeffer's report.

Following this, in 1808, one Gazzeri "made some attempts to utilize the boracic acid," while Mascagni, in 1812, got a patent from the great Napoleon, under which he thought to monopolize the business. Having too many other matters to attend to, however, he turned the patent over to one Fossi, who, in 1818, "exhibited white glass in Florence." Meantime, Gazzeri had been working away, and, with one Brouzet, employed an engineer named Ciasehi to construct artificial pools around the dry ones.

Thereupon occurred the first tragedy in the borax business. Signor Ciasehi, while overseeing some workmen, fell into a chasm. He was dragged out terribly scalded, and, after lingering a few days in excruciating pain, he died.

After working three years, "Gazzeri and Brouzet, with great difficulty, managed to export to France three tons and 5½ cwt. of very impure crude boracic acid," the product of the pools during the 9½ months ending April 1, 1818.

In the same year, M. Francois Lardarel, a Frenchman, then living in Tuscany, went into the boracic acid business in the region under consideration. Although his profits were small, he kept his establishment running until 1827, when he hit on an improvement that made him rich. He had, during those nine years, been using the woody productions of the regions for fuel, to evaporate the acid-charged water so that the acid would crystallize out. It was very much like using sage-brush and grease-wood on the deserts of Nevada and California for fuel for producing borax. It was so very expensive that no profit worth mentioning was left, especially after the brush was cleared away for a considerable distance. But, in 1827, Lardarel hit upon a plan for utilizing the heated gases from the dry holes, and the problem was solved. The production rapidly increased, new uses for the crystal were found, and the Lardarel family became great in the land, and remains so to this day.

The plant, as he arranged it, is the simplest imaginable, and apparently very effective. A series of vats or pools was constructed in line down a hill-side, where the vents for subterranean gases were numerous. Each vat had its jet of gases, and the vats had to be regulated in size to suit the power of the jets. Water was conducted from a spring into the upper vat until it was filled. The water boiled at a great rate there for twenty-four hours, and was then run off into the next vat, while the first was refilled. So the water ran on from vat to vat, until in the last it was found to have, along with a lot of mud and various sulphates, from 1½ to 2 per cent. of boracic acid.

From the last boiling pool or vat the liquor was run into a settling vat, where about all the impurities were precipitated, after which the liquor was conducted into a series of shallow pans, heated by means of steam, formed by letting water into a dry vent that had been covered over with a dome. The gas from the vent turned the water to steam, and the steam ran through a pipe to the evaporating pans.

When the liquor had been reduced to the consistency of syrup, it was conveyed to barrels, where the boracic acid crystallized out. Then the remaining liquor was drawn off, to be used over, and the crystals were dried and shipped to Liverpool. A ton of boracic acid is said to cost but $73 in these days, the cost of labor being very low and the natural advantages great, for the old process is still in use.

A double borate of lime was found near Iquiqui, in South America, in 1836. Large deposits have since been found in Ascotan, in the province of Potosi, Bolivia. This borate is shipped to Autofagasta by rail, and made into borax and boracic acid. There are other smaller deposits in that region, and the product of recent years is estimated at 4,500 tons of borates, equal to 3,000 tons of borax.

Along about the time that the great discoveries of borates were made in Nevada, a foreigner, prospecting in Asia Minor, along Tchinar-sav, a branch of the Rhyndacus River, forty miles from Panderma, on the Sea of Marmora, found a cropping that looked like "a snow-white, fine-grained marble." It existed in closely-packed nodules of very irregular size and shape, and of all weights up to a ton. It was assayed and found to be a compound of boracic acid, lime, and water—a borate of lime rich in the acid. The discoverers worked it for a long time under the pretense that it was plaster of paris, so escaping the payment of duty to the Turks. They found their market in England. The government "was apprised of these irregularities," and "energetic measures to correct them" were taken—Turkish fashion. The truth probably is that the English merchants who were buying the product found it valuable, and determined to jump the claim. So they "gave it away" to the Turk. Thereupon the Turk "granted a comprehensive concession" to a British company, and they have since worked the deposit. "Labor is very cheap and abundant—Turks, Armenians, Greeks, Circassians, Tartars, and Italians being obtainable from the neighboring villages. The borate being new at the time of discovery, was called pandermite, after the port of shipment."

CHAPTER XIX.

CURIOUS FACTS ABOUT BORAX.

AMONG the newspaper writers who have made reputations on the Pacific Coast is one who signs his stories Dan de Quille. Dan tells interesting stories always, and there is reason to believe they are founded more or less upon facts. Once upon a time he turned his pen loose on Death Valley, and it was a weird yarn that followed. There was an outfit of prospectors, and they feared neither Piutes nor grizzlies, deserts nor Death Valley. They were in search of the Gunsight lead and had it almost within their grasp, when something too awful for even the nerves of such men was found. They were crossing the dreaded white marsh in Death Valley, and were as cheerful as linnets, in spite of boiling mercury and the omnipresent horned rattlesnakes, when a red object very unexpectedly hove in view on the port bow, so to speak, and the hardy prospectors felt bound to bear down upon it and investigate. They approached at first with only moderate curiosity, but very quickly this was turned to wonder, for the object assumed the human form. Then they drew nigh, and behold! It was the body of a woman in red calico dress and a pink poke-bonnet, with every form and feature well-nigh intact, even to the look of anguish and distress on her face which had marked her last moments. It was the body of one of the old emigrant party, who, in

THE DRYING FLOORS.

perishing, gave the valley its name. The spectacle was too gruesome; one look, and the hardy prospectors fled in horror.

As was said, Dan de Quille commonly has a basis of fact for his stories. There is a basis of fact for this one. There is a Death Valley, and there was an emigrant party. Further than that, there is a very good preservative of flesh of all sorts in the marsh of Death Valley. The meat-packers of Chicago, St. Louis, Kansas City, and Omaha began using Death Valley borax some years ago, to preserve their merchandise from putrefaction, and, during the year 1891, used between 3,000,000 and 4,000,000 pounds of the article for that purpose.

This calls attention to a number of curious and, one may say, almost contradictory facts about borax. It preserves meat, according to the *United States Dispensatory*, "by destroying the microscopic vegetable and animal organism upon which fermentation and putrefaction depend." Now, although borax destroys such forms of life as those, it can be applied to wounds as an antiseptic without producing any corrosive or dangerous effects, and is used for various kinds of sore throat successfully.

The packers of meat buy their borax at about 8 cents a pound. Sportsmen buy it, as told elsewhere, for the same use—to preserve game—but, getting it under another name, pay something like $2 a pound. The wily drug-maker has compounded a combination of borax and wind—advertising, so to speak—and the sportsman, finding the patent stuff to work well, pays the price, although a box of powdered borax of precisely the same efficiency could be had in a grocery store for not to exceed 15 cents. In like manner it is used to preserve milk from turning sour. People know that milk taken from the cows in Orange County, New York, in the morning, even in the hot days of August, does not reach New York City until the next morning, and that it is still on sale sweet and good in the New York groceries, in spite of all the pounding around and the heat, thirty-six hours after milking time. But mighty few people know that a pinch of borax to the pint kept it so.

The sportsman is not the only one who buys borax at an enormous price. Within recent years a variety of powders, pastes, and solutions have been put on the market for use on the teeth. They are fragrant and refreshing, and sell for 25 cents a bottle or package. A bottle of powder holds about an ounce, and it is a bottle of powdered and colored borax, with a bit of flavor, that costs the man who puts it up possibly 50 or 60 cents a pound.

And there is another toilet use where it is wholly undisguised, or at most is but delicately scented. It is a very popular powder for whitening the faces of ladies who are too much tanned, or have faded in some way. It has the advantage over other bleaches of softening the skin, purifying it, and removing blotches, pimples, and tan, but it is just as good mixed with chalk, if in a manilla or straw box, at 15 cents a pound, as it is in fancy bottles at $5.

In fact, the number of places where borax confronts the housekeeper unexpectedly or in unknown form is very great. The most delicate of Sevres porcelain has borax in it. So, too, has the common table-ware made in New Jersey and elsewhere. One grasps borax when he turns the common white door-knob, and he sees it when he looks at the recently developed granite ware. The woman with a calico dress has borax in the colors used in printing. The man with a stiff hat wears borax in the shape of a varnish that makes the hat stiff.

Nearly every village boy has, at one time and another, seen the blacksmith sprinkle a white powder on the ends of pieces of iron which he was soldering together. The blacksmith said the powder was borax. The use of borax in the blacksmith's fire suggests two other uses of the substance. The safe-makers put it between the inner and outer linings of fire-proof safes, but a more important, though scarcely known, use of it is as a fire-proofing material. Added to lime, in the proportion of two or three pounds to the barrel, it makes plastering practically fire-proof. It is already used in the manufacture of fire-proof paint, and by the manufacturers of fine dress-goods, which they wish to render fire-proof.

It is a fact, interesting to housekeepers, that the most effective insect destroyer is the plain powdered borax. Indeed, a very large number of the insect powders are made of borax, colored to hide its real character from the purchaser. Its use in the laundry, beginning with the days when the Dutch gained fame for their white linens, is so well known (or should be), that only mention need be made of it, while every barber uses a solution of it for shampooing his customers.

There are in all more than fifty different uses of borax, but, because until within recent years it was a high-priced drug instead of a cheap salt, the knowledge of its uses has not been widespread among ordinary people.

In 1707, the price of tincal (i. e., crude borax) in London was £9 5s. a cwt., or about $1,000 per ton. A price-current for July, 1756, gives the price at £280 per ton.

As originally imported into this country the cost was not less than $1 per pound. At the time of the first discovery of borax in California, the wholesale price in New York was about 50 cents per pound. The Pacific Coast production brought this price down to 19 cents at once, and as low as 10 cents by 1873, with a subsequent advance to 15 cents. The production has since then doubled, and the

SACKING BORAX FOR MARKET.

price has fallen to about 8 cents in car-load lots, and the supply is such as to warrant the confidence that it will remain a cheap commodity in the future.

From the various estimates as to the total annual production of borax throughout the world, the following is compiled as probably nearly accurate:

Asia Minor	9,000 tons.
Thibet	2,000 "
Italy	3,000 "
Chile and Bolivia	3,000 "
California and Nevada	6,000 "
Total	23,000 tons.

The largest sources, Asia Minor and the United States, have been developed within the last twenty years, showing that the use of borax has increased within that time nearly 300 per cent.

When borax once enters the family as an article of common use, it soon becomes a necessity, and is as fixed a commodity for regular supply as salt, soap, or sugar. The variety of domestic needs it is fitted to meet renders it a welcome addition to the household.

It Tastes Good

One reason why *Scott's Emulsion* of Pure Norwegian Cod Liver Oil and Hypophosphites of Lime and Soda has had such a large sale is because it is "As palatable as milk;" but the best reason is that its curative properties are unequaled. It cures the cough, supplies the waste of tissues, produces flesh, and builds up the entire system.

Scott's Emulsion cures Coughs, Colds, Consumption, Scrofula, General Debility, and all Anæmic and Wasting Diseases. It prevents wasting in children. ***It is palatable as milk.*** Scott's Emulsion is put up in salmon-colored wrappers. ***Be sure you get the genuine.*** It is prepared only by Scott & Bowne, New York, and sold by all druggists.

Scott's Emulsion

Made in the USA
Monee, IL
04 November 2026

67381756R00134